GAS CLOSED SYSTEM CYCLES

GAS CLOSED SYSTEM CYCLES

CHIH WU

Nova Science Publishers, Inc.
New York

Copyright © 2009 by Nova Science Publishers, Inc.

All rights reserved. No part of this book may be reproduced, stored in a retrieval system or transmitted in any form or by any means: electronic, electrostatic, magnetic, tape, mechanical photocopying, recording or otherwise without the written permission of the Publisher.

For permission to use material from this book please contact us:
Telephone 631-231-7269; Fax 631-231-8175
Web Site: http://www.novapublishers.com

NOTICE TO THE READER
The Publisher has taken reasonable care in the preparation of this book, but makes no expressed or implied warranty of any kind and assumes no responsibility for any errors or omissions. No liability is assumed for incidental or consequential damages in connection with or arising out of information contained in this book. The Publisher shall not be liable for any special, consequential, or exemplary damages resulting, in whole or in part, from the readers' use of, or reliance upon, this material.

Independent verification should be sought for any data, advice or recommendations contained in this book. In addition, no responsibility is assumed by the publisher for any injury and/or damage to persons or property arising from any methods, products, instructions, ideas or otherwise contained in this publication.

This publication is designed to provide accurate and authoritative information with regard to the subject matter covered herein. It is sold with the clear understanding that the Publisher is not engaged in rendering legal or any other professional services. If legal or any other expert assistance is required, the services of a competent person should be sought. FROM A DECLARATION OF PARTICIPANTS JOINTLY ADOPTED BY A COMMITTEE OF THE AMERICAN BAR ASSOCIATION AND A COMMITTEE OF PUBLISHERS.

LIBRARY OF CONGRESS CATALOGING-IN-PUBLICATION DATA
Available upon request

ISBN: 978-1-60741-058-4

Published by Nova Science Publishers, Inc. ✢ New York

CONTENTS

Preface		vii
Chapter 1	Otto Cycle	1
Chapter 1A	Wankel Engine	17
Chapter 2	Diesel Cycle	19
Chapter 3	Atkinson Cycle	33
Chapter 4	Dual Cycle	37
Chapter 5	Lenoir Cycle	43
Chapter 6	Stirling Cycle	49
Chapter 7	Miller Cycle	57
Chapter 8	Wicks Cycle	63
Chapter 9	Rallis Cycle	67
Chapter 10	Design Examples	75
Chapter 11	Summary	93
References		95
Index		977

PREFACE

Heat engines that use gases as the working fluid in a closed system model were discussed in this book. Otto cycle, Diesel, Miller, and Dual cycle are internal combustion engines. Stirling cycle is an external combustion engine. The Otto cycle is a spark-ignition reciprocating engine made of an isentropic compression process, a constant volume combustion process, an isentropic expansion process, and a constant volume cooling process. The thermal efficiency of the Otto cycle depends on its compression ratio. The compression ratio is defined as r=Vmax/Vmin. The Otto cycle efficiency is limited by the compression ratio because of the engine knock problem. The Diesel cycle is a compression-ignition reciprocating engine made of an isentropic compression process, a constant pressure combustion process, an isentropic expansion process, and a constant volume cooling process. The thermal efficiency of the Otto cycle depends on its compression ratio and cut-off ratio. The compression ratio is defined as r=Vmax/Vmin. The cut-off ratio is defined as rcutoff=Vcombustion off/Vmin. The Dual cycle involves two heat addition processes, one at constant volume and one at constant pressure. It behaves more like an actual cycle than either Otto or Diesel cycle. The Lenoir cycle was the first commercially successful internal combustion engine. The Stirling cycle and Wicks cycle are attempt to achieve the Carnot efficiency. The Miller cycle uses variable valve timing for compression ratio control to improve the performance of internal combustion engines.

Chapter 1

OTTO CYCLE

A four stroke internal combustion engine was built by a German engineer, Nicholas Otto, in 1876. The cycle patterned after his design is called the *Otto cycle*. It is the most widely used internal combustion heat engine in automobiles.

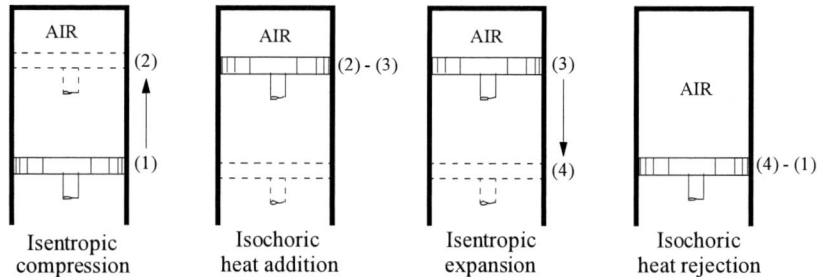

Figure 9.1.1. Otto cycle.

The piston in a four stroke internal combustion engine executes four complete strokes as the crankshaft completes two revolution per cycle as shown in Figure 9.1.1. On the intake stroke, the intake valve is open and the piston moves downward in the cylinder, drawing in a premixed charge of gasoline and air until the piston reaches its lowest point of the stroke called *bottom dead center* (BDC). During the compression stroke the intake valve closes and the piston moves toward the top of the cylinder, compressing the fuel-air mixture. As the piston approaches the top of the cylinder called *top dead center* (TDC), the spark plug is energized and the mixture ignites, creating an increase in the temperature and pressure of the gas. During the expansion stroke the piston is forced down by the high pressure gas, producing a useful work output. The cycle is then completed when the exhaust valve opens and the piston moves toward the top of the cylinder, expelling the products of combustion.

The thermodynamic analysis of an actual Otto cycle is complicated. To simplify the analysis, we consider an ideal Otto cycle composed entirely of internally reversible processes. In the Otto cycle analysis, a closed piston-cylinder assembly is used as a control mass system.

The cycle is made of the following four processes:

1-2 isentropic compression
2-3 constant volume heat addition
3-4 isentropic expansion
4-1 constant volume heat removing

The p-v and T-s process diagrams for the ideal two-stroke Otto cycle are illustrated in Figure 9.1.2.

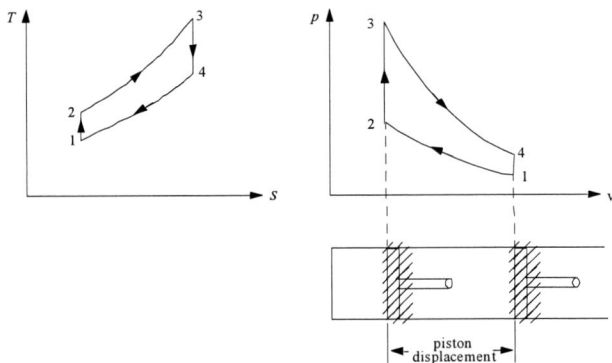

Figure 9.1.2. Otto cycle p-v and T-s diagrams.

Applying the First law and Second law of thermodynamics of the closed system to each of the four processes of the cycle yields:

$$W_{12} = \int p dV \tag{9.1.1}$$

$$Q_{12} - W_{12} = m(u_2 - u_1), \; Q_{12} = 0 \tag{9.1.2}$$

$$W_{23} = \int pdV = 0 \tag{9.1.3}$$

$$Q_{23} - 0 = m(u_3 - u_2) \tag{9.1.4}$$

$$W_{34} = \int pdV \tag{9.1.5}$$

$$Q_{34} - W_{34} = m(u_4 - u_3), \ Q_{34} = 0 \tag{9.1.6}$$

$$W_{41} = \int pdV = 0 \tag{9.1.7}$$

and

$$Q_{41} - 0 = m(u_1 - u_4) \tag{9.1.8}$$

The net work (W_{net}), which is also equal to net heat (Q_{net}), is

$$W_{net} = W_{12} + W_{34} = Q_{net} = Q_{23} + Q_{41} \tag{9.1.9}$$

The thermal efficiency of the cycle is

$$\eta = W_{net}/Q_{23} = Q_{net}/Q_{23} = 1 - Q_{41}/Q_{23} = 1 - (u_4 - u_1)/(u_3 - u_2) \tag{9.1.10}$$

This expression for thermal efficiency of an ideal Otto cycle can be simplified if air is assumed to be the working fluid with constant specific heats. Equation (9.1.10) is reduced to:

$$\eta = 1 - (T_4 - T_1)/(T_3 - T_2) = 1 - (r)^{1-k} \tag{9.1.11}$$

where r is the *compression ratio* for the engine defined by the equation

$$r = V_1/V_2 \tag{9.1.12}$$

The compression ratio is the ratio of the cylinder volume at the beginning of the compression process (BDC) to the cylinder volume at the end of the compression process (TDC).

Equation (9.1.11) shows that the thermal efficiency of the Otto cycle is only a function of the compression ratio of the engine. Therefore, any engine design that increases the compression ratio should result in an increased engine efficiency. However, the compression ratio cannot be increased indefinitely. As the compression ratio increases, the temperature of the working fluid also increases during the compression process. Eventually, a temperature is reached that is sufficiently high to ignite the air-fuel mixture prematurely without the presence of a spark. This condition causes the engine to produce a noise called knock. The presence of *engine knock* places a barrier on the upper limit of Otto engine compression ratios. To reduce engine knock problem of a high compression ratio Otto cycle, one must use gasoline with higher octane rating. In general, the higher the octane rating number of gasoline, the higher the resistance of engine knock.

One way to simplify the calculation of the net work of the cycle and to provide a comparative measure of the performance of an Otto heat engine is to introduce the concept of the mean effective pressure. The *mean effective pressure* (MEP) is the average pressure of the cycle. The net work of the cycle is equal to the mean effective pressure multiplied by the displacement volume of the cylinder. That is

MEP = (cycle net work)/(cylinder displacement volume) = $W_{net}/(V_1-V_2)$

The engine with the larger MEP value of two engines of equal cylinder displacement volume would be the better one, because it would produce a greater net work output.

The intake and exhaust in the two-stroke Otto cycle occur instantaneously. In order to make the intake and exhaust processes much better, a four-stroke Otto cycle is commonly used as shown in Figure 9.1.3.

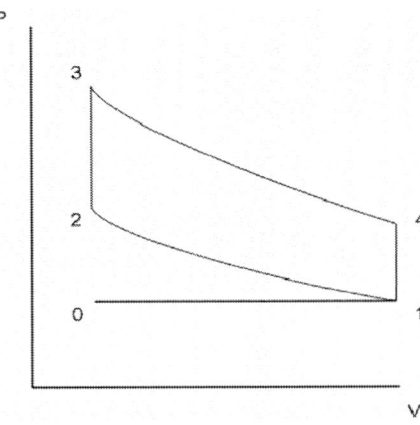

Figure 9.1.3. Four-stroke Otto cycle.

When the piston reaches BDC, the exhaust valve is opened, reducing cylinder pressure to the initial pressure, with a corresponding decrease of temperature (process 4-1). Finally to complete the four-stroke cycle, with the exhaust valve open, the piston is pushed upward (process 1-0), clearing the cylinder of the combustion gases. For each complete cycle of the four-stroke cycle, there are four strokes and hence two crankshaft revolutions. The power developed by the engine is give by

$\dot{W} = W_{net}(N/2)$

where $\dot{W}$ is the power output, W_{net} is the net engine work output per cycle, and N is the crankshaft revolutions per unit time, respectively.

The Otto cycle can operate either on a two-stroke or a four-stroke cycle.

The advantage of the two-stroke Otto cycle is that it provides twice as many power strokes as the four-stroke cycle per cylinder per crank shaft revolution. However, the processes of intake and exhaust scavenging are not as efficient as those with a four-stroke cycle. It is possible to lose some of the fresh fuel-air mixture out of the exhaust prior to combustion, and it is also possible for an appreciable fraction of the burned gases to remain in the cylinder. For these reasons, the actual power output of the two-stroke cycle is certainly not twice as great as might be predicted from the number of power strokes per revolution. Also, the poor intake and scavenging efficiency of the two-stroke cycle leads to a worsening in fuel economy, compared to a four-stroke cycle. Furthermore, with the crankcase of the two-stroke cycle used for compressing the incoming charge, it is not available for lubrication. Therefore, the two-stroke engine cannot be lubricated as easily as the four-stroke engine. Oil must be mixed with the fuel in the two-stroke engine to achieve adequate lubrication. For all these reasons, the two-stroke Otto engine has only a limited application in which fuel economy and pollution are not primary factors.

EXAMPLE 9.1.1.

An engine operates on the Otto cycle and has a compression ratio of 8. Fresh air enters the engine at 27°C and 100 kPa. The amount of heat addition is 700 kJ/kg. The amount of air mass in the cylinder is 0.01 kg. Determine the pressure and temperature at the end of the combustion, the pressure and temperature at the end of the expansion, MEP, efficiency and work output per kilogram of air. Show the cycle on T-s diagram. Plot the sensitivity diagram of cycle efficiency vs compression ratio.

To solve this problem by CyclePad, we take the following steps:

1. Build
   (A) Take a compression device, a combustion chamber, an expander and a cooler from the closed system inventory shop and connect the four devices to form the Otto cycle as shown in Figure 9.1.2.
   (B) Switch to analysis mode.
2. Analysis
   (A) Assume a process for each of the four processes: (a) compression device as adiabatic and isentropic, (b) combustion as isochoric, (c) expander as adiabatic and isentropic, and (d) cooler as isochoric.
   (B) Input the given information: (a) working fluid is air, (b) the inlet pressure and temperature of the compression device are 100 kPa and 27°C, (c) the compression ratio of the compression device is 8, (d) the heat addition is 700 kJ/kg in the combustion chamber, and (e) m=0.01 kg.
3. Display results
   (A) Display the T-s diagram and cycle properties results as shown in Figure E9.1.1.a and b. The cycle is a heat engine. The answers are: p=4441 kPa and T=1393°C (the pressure and temperature at the end of the combustion), p=241.6 kPa and T=452.1°C (the pressure and temperature at the end of the expansion), MEP=525.0 kPa, η=56.47% and Wnet=3.95 kJ, and
   (B) Display the sensitivity diagram of cycle efficiency vs compression ratio as shown in Figure E9.1.1.c.

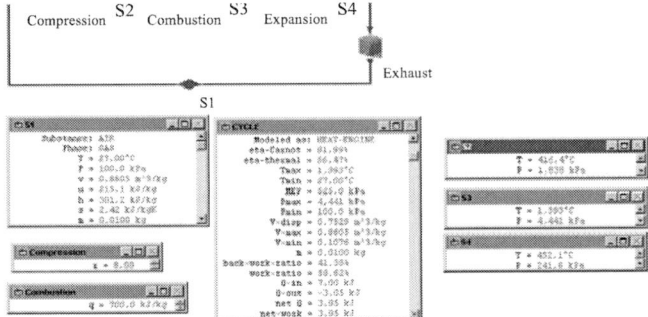

Figure E9.1.1a. Otto cycle

Otto Cycle

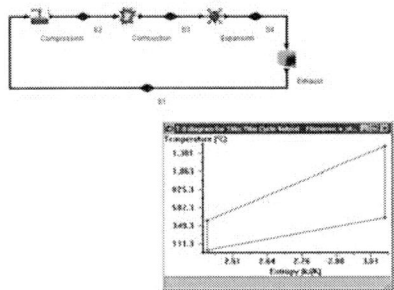

Figure E9.1.1b. Otto cycle T-s diagram

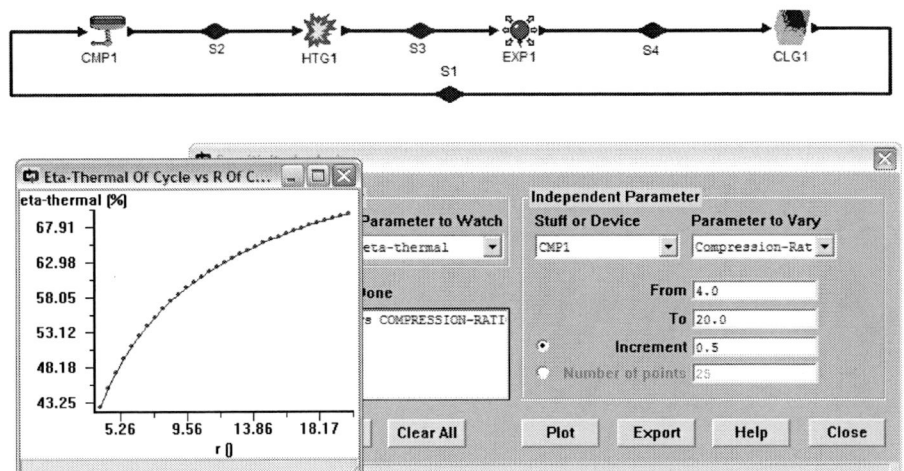

Figure E9.1.1c. Otto cycle sensitivity analysis

Comment: Efficiency increases as compression ratio increases.

EXAMPLE 9.1.2.

The compression ratio in an Otto cycle is 8. If the air before compression (state 1) is at 60°F and 14.7 psia and 800 Btu/lbm is added to the cycle and the mass of air contained in the cylinder is 0.025 lbm, Calculate (a) temperature and pressure at each point of the cycle, (b) the heat must be removed, (c) the thermal efficiency, and (d) the MEP of the cycle.

To solve this problem by CyclePad, we take the following steps:

1. Build
   (A) Take a compression device, a combustion chamber, an expander and a cooler from the closed system inventory shop and connect the four devices to form the Otto cycle.
   (B) Switch to analysis mode.
2. Analysis
   (A) Assume a process for each of the four processes: (a) compression device as isentropic, (b) combustion as isochoric, (c) expander as isentropic, and (d) cooler as isochoric.
   (B) Input the given information: (a) working fluid is air, (b) the inlet pressure and temperature of the compression device are 60°F and 14.7 psia, (c) the compression ratio of the compression device is 8, and (d) the heat addition is 800 Btu/lbm in the combustion chamber, and (e) the mass of air is 0.025 lbm.
3. Display results
   (A) Display the T-s diagram and cycle properties results as shown in Figure E9.1.2.a and b. The cycle is a heat engine. The answers are T_2=734.2 °F, p_2=270.2 psia, T_3=5,407 °F, p_3=1,328 psia, T_4=2,094 °F, p_4=72.24 psia, η=56.47%, Q_{41}=-8.71 Btu, and MEP=213.3 psia.
   (B) Display the T-s diagram.

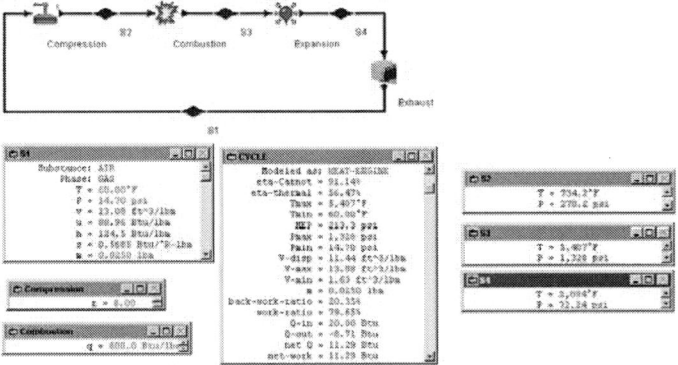

Figure E9.1.2a. Otto cycle.

The power output of the Otto cycle can be increased by turbo-charging the air before it enters the cylinder in the Otto engine. Since the inlet air density is increased due to higher inlet air pressure, the mass of air in the cylinder is increased. Turbo-charging raises the inlet air pressure of the engine above atmospheric pressure and raise the power output of the engine, but it may not

Otto Cycle

improve the efficiency of the cycle. The schematic diagram of the Otto cycle with turbo-charging is illustrated in Figure 9.1.3. Example 9.1.3 and Example 9.1.4 show the power increase due to turbo-charging.

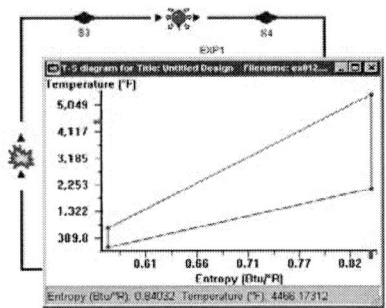

Figure E9.1.2b. Otto cycle T-s diagram plot.

Figure 9.1.3. Otto engine with turbo-charging.

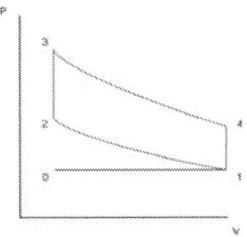

Figure E9.1.4. Otto engine without turbo-charging.

EXAMPLE 9.1.3

Determine the heat supplied, work output, MEP, and thermal efficiency of an ideal Otto cycle with a compression ratio of 10. The highest temperature of the cycle is 3000°F. The volume of the cylinder before compression is 0.1 ft^3. What is the mass of air in the cylinder? The atmosphere conditions are 14.7 psia and 70°F.

To solve this problem, we build the cycle. Then (A) Assume isentropic for compression process 1-2, isentropic for compression process 2-3, isochoric for the heating process 3-4, isentropic for expansion process 4-5, and isochoric for the cooling process 5-6; (B) input p_1 =14.7 psia, T_1=70°F, V_1=0.1 ft^3; p_2=14.7 psia, T_2=70°F, V_2=0.1 ft^3 (no turbo-charger); compression ratio=10, T_4=3000°F, p_6=14.7 psia, and T_6=70°F; and (C) display results. The results are: W_{12}=-0 Btu, W_{23}=-1.03 Btu, Q_{34}=2.73 Btu, W_{45}=2.67 Btu, W_{net}=1.65 Btu, Q_{56}=-1.09 Btu, MEP=98.80 psia, η=60.19%, and m=0.0075 lbm.

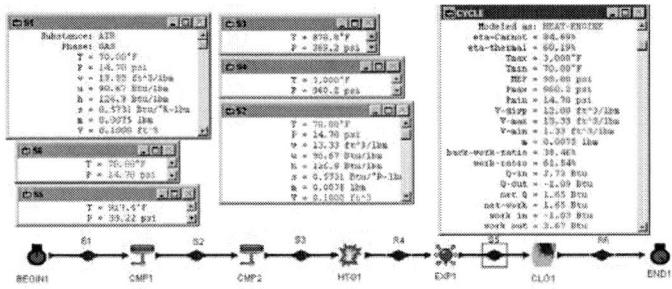

Figure E9.1.3. Otto engine without turbo-charging.

EXAMPLE 9.1.4

Determine the heat supplied, work output, MEP, and thermal efficiency of an ideal Otto cycle with a turbo-charger which compresses air to 20 psia and compression ratio of 10. The highest temperature of the cycle is 3000°F. The volume of the cylinder before compression is 0.1 ft^3. What is the mass of air in the cylinder? The atmosphere conditions are 14.7 psia and 70°F.

To solve this problem, we build the cycle as shown in Figure 9.1.3. Then (A) Assume isentropic for compression process 1-2, isentropic for compression process 2-3, isochoric for the heating process 3-4, isentropic for expansion process 4-5, and isochoric for the cooling process 5-6; (B) input p_1 =14.7 psia, T_1=70°F; p_2=20 psia, V_2=0.1 ft^3 (with turbo-charger); compression ratio=10, T_4=3000°F, p_6=14.7 psia, and T_6=70°F; and (C) display results. The results are: W_{13}=-1.48 Btu, Q_{34}=3.21 Btu, W_{45}=3.52 Btu, W_{net}=2.04 Btu, Q_{56}=-1.17 Btu, MEP=96.20 psia, η=63.54%, and m=0.0093 lbm.

Modern car Otto engine designs are affected by environmental constrains as well as desires to increase gas mileage. Recent design improvements include the use of four valves per cylinder to reduce the restriction to air flow into and out of the cylinder, turbo-charges to increase the air and fuel flow to each cylinder, catalytic converters to aid the combustion of unburned hydrocarbons that are expelled by the engine, among others.

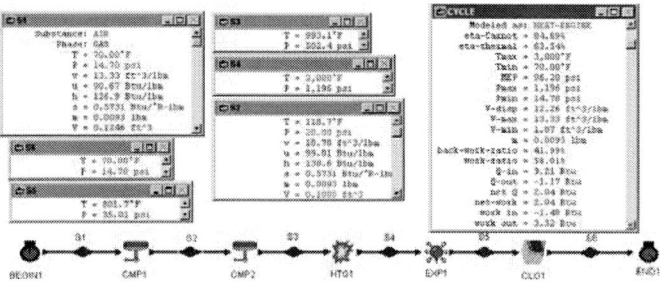

Figure E9.1.4. Otto engine with turbo-charging.

HOMEWORK 9.1. OTTO CYCLE

1. Do Otto heat engines operate on a closed system or an open system? Why?
2. List the four processes in the Otto cycle.
3. How does the two-stroke Otto cycle differ from the four-stroke Otto cycle?
4. What is the ratio of the number of power strokes in the two-stroke cycle divided by the number of power strokes in the four-stroke cycle at a given value of engine revolutions per minute?
5. On what single factor does the efficiency of the Otto cycle depend?
6. What is compression ratio of an Otto cycle? How does it affect the thermal efficiency of the cycle?
7. What limits the practical realization of higher efficiencies in the Otto cycle?
8. Do you know the compression ratio of your car? Is there any limit to an Otto cycle? Why?
9. Which area represents cycle net work of an Otto cycle plotted on a T-s diagram and p-v diagram?
10. How do you define MEP (mean effect pressure)? Can MEP of a car in operation lower than the atmospheric pressure?
11. What is engine knock? What cause the engine knock problem?

12. Do you get a better performance using premier gasoline (Octane number 93) for your compact car?
13. How does the modern Otto cycle achieve higher power output without the use of higher compression ratio?
14. How does the clearance volume affect the efficiency of the Otto cycle?
15. In an ideal Otto cycle, indicate whether the following statements are true or false:
    (A) All the processes are internally reversible.
    (B) Cycle efficiency increases with the maximum temperature.
    (C) There is a constant ratio between the work and the mean effective pressure.
    (D) The gas temperature after compression is higher than after expansion.
    (E) Cycle efficiency depends on the temperature ratio during compression.
16. Sketch T-s and p-v diagrams for the Otto cycle.
17. For an Otto cycle, plot the cycle efficiency as a function of compression ratio from 4 to 16.
18. For an Otto cycle, plot the MEP as a function of compression ratio from 4 to 16.
19. Does the initial state of the compression process have any influence on the Otto cycle efficiency?
20. What is the initial state of the compression process of your car?
21. How many parameters do you need to know to completely describe the Otto cycle?
22. The inlet and exhaust flow processes are not included in the analysis of the Otto cycle. How do these processes affect the Otto cycle performance?
23. How does the Otto cycle efficiency compare to the Carnot cycle efficiency when operating between the same temperature range?
24. As a car gets older, will its compression ratio change? How about the MEP?
25. An engine operates on an Otto cycle with a compression ratio of 8. At the beginning of the isentropic compression process, the volume, pressure and the temperature of the air are 0.01 m^3, 110 kPa and 50°C. At the end of the combustion process, the temperature is 900°C. Find (A) the temperature at the remaining two states of the Otto cycle, (B) the pressure of the gas at the end of the combustion process, (C) heat added per unit mass to the engine in the combustion chamber, (D) heat removed per unit mass from the engine to the environment, (E) the compression work per

Otto Cycle

unit mass added, (F) the expansion work per unit mass done, (G) MEP, and (H) thermal cycle efficiency.

ANSWER: (A) 469.3°C, 237.5°C, (B) 2022 kPa, (C) 3.67 kJ/kg, (D) -1.60 kJ/kg, (E) -300.5 kJ/kg, (F) 474.8 kJ/kg, (G) 236.6 kPa, (H) 56.47%.

26. An ideal Otto Cycle with air as the working fluid has a compression ratio of 9. At the beginning of the compression process, the air is at 290 K and 90 kPa. The peak temperature in the cycle is 1800 K. determine: (A) the pressure and temperature at the end of the expansion process (power stroke), (B) the heat per unit mass added in kJ/kg during the combustion process, (C) net work, (D) thermal efficiency of the cycle, and (E) mean effective pressure in kPa.

ANSWER: (A) 232.0 kPa, 747.4 K, (B) 789.6 kJ/kg, (C) 461.7 kJ/kg, (D) 58.48%, (E) 562.3 kPa.

27. An ideal Otto engine receives air at 15 psia, 0.01 ft^3 and 65°F. The maximum cycle temperature is 3465°F and the compression ratio of the engine is 7.5. Determine (A) the work added during the compression process, (B) the heat added to the air during the heating process, (C) the work done during the expansion process, (D) the heat removed from the air during the cooling process, and (E) the thermal efficiency of the cycle.

ANSWER: (A) -0.1625 Btu, (B) 0.3638 Btu, (C) 0.2873 Btu, (D) -0.1625 Btu, (E) 55.33%.

28. An ideal Otto engine receives air at 14.6 psia and 55°F. The maximum cycle temperature is 3460°F and the compression ratio of the engine is 10. Determine (A) the work done per unit mass during the compression process, (B) the heat added per unit mass to the air during the heating process, (C) the work done per unit mass during the expansion process, (D) the heat removed per unit mass from the air during the cooling process, and (E) the thermal efficiency of the cycle.

ANSWER: (A) -133.2 kJ/kg, (B) 449.7 kJ/kg, (C) 403.9 kJ/kg, (D) -179.0 kJ/kg, (E) 60.19%.

29. An ideal Otto engine receives air at 100 kPa and 25°C. Work is performed on the air in order to raise the pressure at the end of the compression process to 1378 kPa. 400 kJ/kg of heat is added to the air during the heating process. Determine (A) the work done during the compression process, (B) the compression ratio, (C) the work done during the expansion process, (D) the heat removed from the air during the cooling process, (E) the MEP (mean effective pressure), and (F) the thermal efficiency of the cycle.

ANSWER: (A) -238.5 kJ/kg, (B) 6.51, (C) 449.4 kJ/kg, (D) -189.0 kJ/kg, (E) 291.6 kPa, (F) 52.74%.

30. At the beginning of the compression process of an air-standard Otto cycle, p=100 kPa, T=290 K and V=0.04 m^3. The maximum temperature in the cycle is 2200 K and the compression ratio is 8. Determine (A) the heat addition, (B) the net work, (C) the thermal efficiency, and (D) the MEP.
ANSWER: (a) 52.89 kJ, (b) 29.87 kJ, (c) 56.47%, (d) 853.3 kPa.

31. An Otto engine operates with a compression ratio of 8.5. The following information is known:
Temperature prior to the compression process: 70°F.
Volume prior to the compression process: 0.05 ft^3
Pressure prior to the compression process: 14.7 psia.
Heat added during the combustion process: 345 Btu/lbm.
Determine: (A) the mass of air in the cylinder, (B) the temperature and pressure at each process endpoint, (C) the compression work and expansion work in Btu/lbm, and (d) the thermal efficiency.
ANSWER: (A) 0.0038 lbm, (B) 787.1°F and 294.1 psia, 2802°F and 769.5 psia, and 926.2°F and 38.46 psia, (C) -122.8 Btu/lbm and 321.2 Btu/lbm, (D) 57.52%.

32. The compression ratio in an Otto cycle is 8. If the air before compression (state 1) is at 80°F and 14.7 psia and 800 Btu/lbm is added to the cycle and the mass of air contained in the cylinder is 0.02 lbm, Find the heat added, heat removed, work added, work produced, net work produced, MEP and efficiency of the cycle.
ANSWER: heat added=16 Btu, heat removed=-6.96 Btu, work added=-2.4 Btu, work produced=11.43 Btu, net work produced=9.04 Btu, MEP=205.4 psia, and efficiency of the cycle=56.47%.

33. The compression ratio in an Otto cycle is 10. If the air before compression (state 1) is at 60°F and 14.7 psia and 800 Btu/lbm is added to the cycle and the mass of air contained in the cylinder is 0.02 lbm, Find the heat added, heat removed, work added, work produced, net work produced, MEP and efficiency of the cycle.
ANSWER: heat added=16 Btu, heat removed=-6.37 Btu, work added=-2.69 Btu, work produced=12.32 Btu, net work produced=9.63 Btu, MEP=221.0 psia, and efficiency of the cycle=60.19%.

34. The compression ratio in an Otto cycle is 16. If the air before compression (state 1) is at 60°F and 14.7 psia and 800 Btu/lbm is added to the cycle and the mass of air contained in the cylinder is 0.02 lbm,

Find the heat added, heat removed, work added, work produced, net work produced, MEP and efficiency of the cycle.
ANSWER: heat added=16 Btu, heat removed=-5.28 Btu, work added=-3.61 Btu, work produced=14.34 Btu, net work produced=10.72 Btu, MEP=236.2 psia, and efficiency of the cycle=67.01%.

35. An Otto engine with a turbo-charger operates with a compression ratio of 8.5. The following information is known:
Temperature prior to the turbo-charging compression process: 70°F.
Pressure prior to the turbo-charging compression process: 14.7 psia.
Pressure after the turbo-charging compression process: 20 psia.
Heat added during the combustion process: 345 Btu/lbm.
Volume after the compression process: 0.05 ft^3
Determine: (A) the mass of air in the cylinder, (B) the temperature and pressure at each process endpoint, (C) the compression work and expansion work in Btu/lbm, and (D) the thermal efficiency.
ANSWER: (A) 0.0033 lbm, (B) 118.7°F and 20.0 psia, 804.4°F and 308.7 psia, 2820°F and 800.9 psia, and 914.4°F and 38.14 psia, (C) -125.7 Btu/lbm and 362.2 Btu/lbm, (D) 58.10%.

36. An ideal Otto Cycle with a turbo-charger using air as the working fluid has a compression ratio of 9. The volume of the cylinder is 0.01 m^3. At the beginning of the turbo-charging compression process, the air is at 290 K and 90 kPa. The air pressure is 150 kPa after the turbo-charging compression process. The peak temperature in the cycle is 1800 K. determine: (A) the pressure and temperature at the end of the expansion process (power stroke), (B) the heat per unit mass added in kJ/kg during the combustion process, (C) net work, (D) thermal efficiency of the cycle, and (E) mean effective pressure in kPa.
ANSWER: (A) 200.5 kPa, 645.9 K, (B) 710.9 kJ/kg, (C) 7.11 kJ/kg, (D) 64.11%, (E) 534.6 kPa.

37. A gasoline engine has a volumetric compression ratio of 9. The state before compression is 290 K, 90 kPa, and the peak cycle temperature is 1800 K. Find the pressure after expansion, work input, work output, net work output, heat input, thermal efficiency, and mean effective pressure of the cycle.
ANSWER: 1951 kPa, -292.7 kJ/kg, 754.4 kJ/kg, 461.7 kJ/kg, 789.6 kJ/kg, 58.48%, 562.3 kPa.

38. A gasoline engine has a volumetric compression ratio of 12. The state before compression is 290 K, 100 kPa, and the peak cycle temperature is 1800 K. Find the pressure after expansion, work input, work output, net

work output, heat input, thermal efficiency, and mean effective pressure of the cycle.
ANSWER: 3242 kPa, -353.7 kJ/kg, 812.6 kJ/kg, 458.9 kJ/kg, 728.5 kJ/kg, 62.99%, 602.1 kPa.

39. A gasoline engine has a volumetric compression ratio of 8 and before compression has air at 280 K and 85 kPa. The combustion generates a peak pressure of 6500 kPa. Find the peak temperature, work input, work output, net work output, heat input, thermal efficiency, and mean effective pressure of the cycle.
ANSWER: 2676 K, -260.4 kJ/kg, 1083 kJ/kg, 822.9 kJ/kg, 1457 kJ/kg, 56.47%, 995.9 kPa.

40. A gasoline engine has a compression ratio of 10:1 with 4 cylinders of total displacement 2.3 L. The inlet state is 280 K and 70 kPa. The fuel adds 1800 kJ/kg of heat in the combustion process. Find the work input, work output, net work output, heat input, thermal efficiency, and mean effective pressure of the cycle.
ANSWER: -0.6085 kJ, 2.78 kJ, 2.17 kJ, 3.61 kJ, 60.19%, 1050 kPa.

41. A gasoline engine has a compression ratio of 12:1 with 4 cylinders of total displacement 2.3 L. The inlet state is 280 K and 100 kPa. The fuel adds 1800 kJ/kg of heat in the combustion process. Find the work input, work output, net work output, heat input, thermal efficiency, and mean effective pressure of the cycle.
ANSWER: -0.9786 kJ, 4.23 kJ, 3.25 kJ, 5.16 kJ, 62.99%, 1541 kPa.

Chapter 1A

WANKEL ENGINE

A spark-ignited internal combustion rotary engine is Wankel engine. In place of the reciprocating motion of the piston, the engine substitutes the rotary motion of an equilateral triangular curved shaped rotor inside a housing to compress and expand the working fluid. As the rotor within the stator (chamber), the volume between the rotor and the stator changes to compress the fuel-air mixture. Since the number of its moving components is less than that of a conventional reciprocating piston engine, the Wankel engine is expected to be more efficient than a conventional reciprocating piston engine.

As shown in Figure 9.1.5, the rotor divides the housing into three volumes. Let us follow volume A as it passes through a cycle. The air-fuel mixture enters the engine in process 1-2 as shown in Figure 9.1.5. As the rotor turns, this volume is sealed off and compressed, corresponding to the compression stroke 2-3. When the volume reaches a minimum (process 3-4, corresponding to TDC), the spark is fired and combustion takes lace. The hot gas then expands and turns the rotor in the power stroke, process 4-5. Finally, in process 5-2-1, the exhaust ports are uncovered to the volume, and the gas are exhausted from the engine. The p-V diagram is exactly the same as that of the Otto cycle.

Note that there are three volumes of gas at various stages of the cycle at a given time. In other words, there are three power strokes per rotor revolution. The output shaft of the engine is geared to run at three times the rotor angular velocity, so that there is one power stroke for each output shaft revolution.

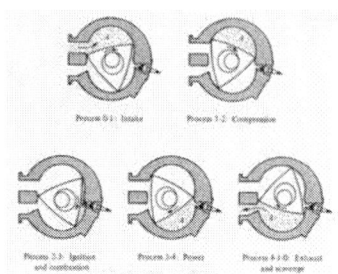

Figure 9.1.5. Wankel engine.

The Wankel engine has a high power to weight ratio, with few parts than that of a conventional Otto piston engine. For example, a six cylinder piston engine with twelve valves and accompanying hardware to control their motion can be replaced by a two rotor rotary engine with no valves. Further the high inertia force of the reciprocating piston and the accompanying noise and vibration are replaced by the smooth and quiet rotary motion of the engine. However, the rotary engine has relatively high emissions of unburned fuel. The problems of reducing emissions to meet standards as well as achieving a competitive fuel economy have prevented widespread use of the engine to date.

HOMEWORK 9.1A. WANKEL CYCLE

1. *Describe the four events of the Wankel engine.*
2. *Describe the operation of the Wankel engine.*
3. *Do you expect the Wankel engine to be more efficient than a conventional reciprocating piston engine? Why?*
4. *Describe the problems that the practical Wankel engine encountered.*

Chapter 2

DIESEL CYCLE

The Diesel cycle was proposed by Rudolf Diesel in the 1890s. The Diesel cycle as shown in Figure 9.2.1 is somewhat similar to the Otto cycle, except that ignition of the fuel-air mixture is caused by spontaneous combustion owing to the high temperature that results from compressing the mixture to a very high pressure. The basic components of the Diesel cycle are the same as the Otto cycle, except that the spark plug is replaced by a fuel injector and the stroke of the piston is lengthened to provide a larger compression ratio.

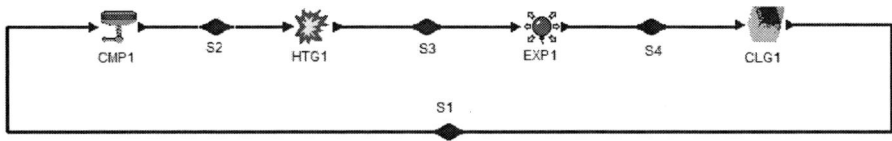

Figure 9.2.1. Diesel cycle.

The Diesel cycle consists of the following four processes:

1-2   isentropic compression
2-3   constant pressure heat addition
3-4   isentropic expansion
4-1   constant volume heat removing

Since the duration of the heat addition process is extended, this process is modeled by a constant pressure process. The p-v and T-s diagram for the Diesel cycle are illustrated in Figure 9.2.2.

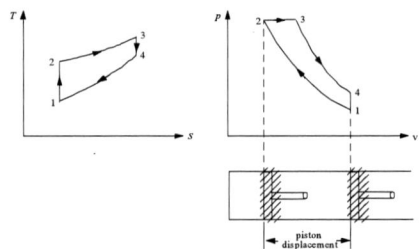

Figure 9.2.2. Diesel cycle p-v and T-s diagrams

Applying the First law and Second law of thermodynamics of the closed system to each of the four processes of the cycle yields:

$$W_{12} = \int pdV \tag{9.2.1}$$

$$Q_{12} - W_{12} = m(u_2 - u_1), \ Q_{12} = 0, \tag{9.2.2}$$

$$W_{23} = \int pdV = m(p_3v_3 - p_2v_2) \tag{9.2.3}$$

$$Q_{23} = m(u_3 - u_2) + W_{23} = m(h_3 - h_2) \tag{9.2.4}$$

$$W_{34} = \int pdV \tag{9.2.5}$$

$$Q_{34} - W_{34} = m(u_4 - u_3), \ Q_{34} = 0 \tag{9.2.6}$$

$$W_{41} = \int pdV = 0 \tag{9.2.7}$$

and

$$Q_{41} - 0 = m(u_1 - u_4). \tag{9.2.8}$$

The net work (W_{net}), which is also equal to net heat (Q_{net}), is

$$W_{net} = W_{12} + W_{23} + W_{34} = Q_{net} = Q_{23} + Q_{41} \tag{9.2.9}$$

The thermal efficiency of the cycle is

$$\eta = W_{net}/Q_{23} = Q_{net}/Q_{23} = 1 - Q_{41}/Q_{23} = 1 - (u_4 - u_1)/(h_3 - h_2) \tag{9.2.10}$$

This expression for thermal efficiency of an ideal Otto cycle can be simplified if air is assumed to be the working fluid with constant specific heats. Equation (9.2.10) is reduced to:

$$\eta = 1 - (T_4 - T_1)/[k(T_3 - T_2)] = 1 - (r)^{1-k}\{[(r_c)^k - 1]/[k(r_c - 1)]\} \tag{9.2.11}$$

where r is the compression ratio, and r_c is the cut-off ratio for the engine defined by the equation

$$r = V_1/V_2 \tag{9.2.12}$$

and

$$r_c = V_3/V_2 \tag{9.2.13}$$

A comparison of the thermal efficiency of the Diesel cycle and the thermal efficiency of the Otto cycle shows that the two thermal cycle efficiencies differ by the quantity in the brackets of equation (9.2.11). This bracket factor is always larger than one, hence the Diesel cycle efficiency is always less than the Otto cycle efficiency operating at the same compression ratio.

Since the fuel is not injected into the cylinder until after the air has been completely compressed in the Diesel cycle, there is no engine knock problem. Therefore the Diesel engine can be designed to operate at much higher compression ratios and less refined fuel than those of the Otto cycle. As a result of the higher compression ratio, Diesel engines are slightly more efficient than Otto engines.

EXAMPLE 9.2.1

A Diesel engine receives air at 27°C and 100 kPa. The compression ratio is 18. The amount of heat addition is 500 kJ/kg. The mass of air contained in the cylinder is 0.0113 kg. Determine (a) the maximum cycle pressure and maximum cycle temperature, (b) the efficiency and work output, and (c) the MEP. Plot the sensitivity diagram of cycle efficiency vs compression ratio.

To solve this problem by CyclePad, we take the following steps:

1. Build

(A) Take a compression device, a combustion chamber, an expander and a cooler from the closed system inventory shop and connect the four devices to form the Diesel cycle.
(B) Switch to analysis mode.
2. Analysis
(A) Assume a process for each of the four processes: (a) compression device as isentropic, (b) combustion as isobaric, (c) expander as isentropic, and (d) cooler as isochoric.
(B) Input the given information: (a) working fluid is air, (b) the inlet pressure and temperature of the compression device are 100 kPa and 27°C, (c) the compression ratio of the compression device is 18, (d) the heat addition is 500 kJ/kg in the combustion chamber, and (e) the mass of air is 0.0113 kg.
3. Display results
(A) Display cycle properties results. The cycle is a heat engine. The answers are T_{max}=1179 °C, p_{max}=5720 kPa, η=65.53%, MEP=403.2 kPa and Wnet=4.26 kJ, and
(B) Display the sensitivity diagram of cycle efficiency vs compression ratio.

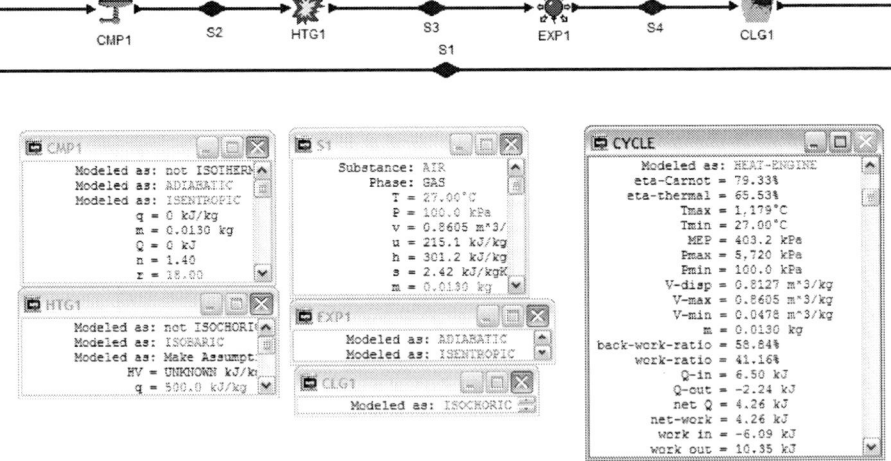

Figure E9.2.1a. Diesel cycle.

Diesel Cycle

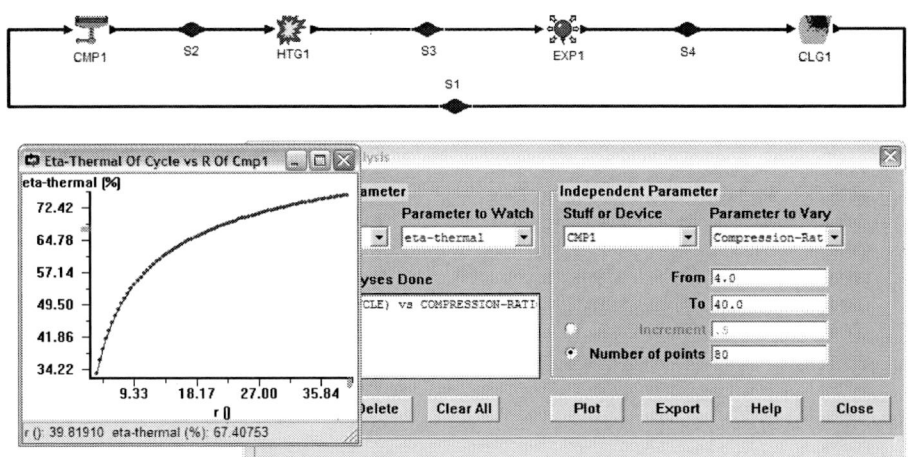

Figure E9.2.1b. Diesel cycle sensitivity analysis.

Comment: Efficiency increases as compression ratio increases.

EXAMPLE 9.2.2

A Diesel engine receives air at 60°F and 14.7 psia. The compression ratio is 16. The amount of heat addition is 800 Btu/lbm. The mass of air contained in the cylinder is 0.02 lbm. Determine the maximum cycle temperature, heat added, heat removed, work added, work produced, net work produced, MEP and efficiency of the cycle.

To solve this problem by CyclePad, we take the following steps:

1. Build
   (A) Take a compression device, a combustion chamber, an expander and a cooler from the closed system inventory shop and connect the four devices to form the Diesel cycle.
   (B) Switch to analysis mode.
2. Analysis
   (A) Assume a process for each of the four processes: (a) compression device as isentropic, (b) combustion as isobaric, (c) expander as isentropic, and (d) cooler as isochoric.
   (B) Input the given information: (a) working fluid is air, (b) the inlet temperature and pressure of the compression device are 60°F and 14.7 psia, (c) the compression ratio of the compression device is 16,

(d) the heat addition is 800 Btu/lbm in the combustion chamber, and
(e) the mass of air is 0.02 lbm.
3. Display results
   (A) Display cycle properties results. The cycle is a heat engine. The answers are T_{max}=4454°F, Q_{add}=16 Btu, Q_{remove}=-6.97 Btu, W_{add}=-3.61 Btu, $W_{expansion}$=12.65 Btu, Wnet=9.03 Btu, MEP=199 psia, and η=56.45%.

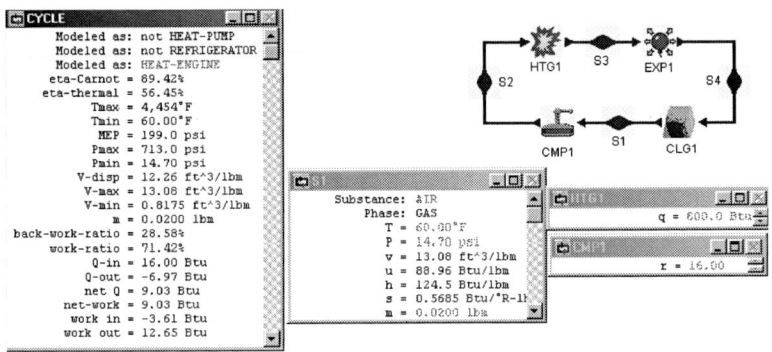

Figure E9.2.2. Diesel cycle.

The power output of the Diesel cycle can be increased by super-charging, turbo-charging and pre-cooling the air before it enters the cylinder in the Diesl engine. The difference between a super-charger and a turbo-charger is the manner in which they are powered. Since the inlet air density is increased due to higher inlet air pressure or lower air temperature, the mass of air in the cylinder is increased. Turbo-charging raises the inlet air pressure of the engine above atmospheric pressure and raise the power output of the engine, but it may not improve the efficiency of the cycle. The schematic diagram of the Diesel cycle with turbo-charging or super-charging is illustrated in Figure 9.2.3. The bottom schematic diagram of Figure 9.2.4 illustrates the Diesel cycle with turbo-charging and pre-cooling. The following three examples (Example 9.2.2, Example 9.2.3, and Example 9.2.4) show the power increase due to super-charging, and pre-cooling and super-charging.

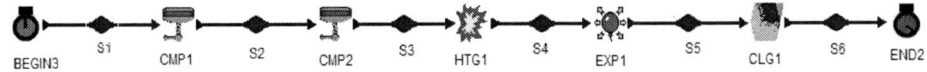

Figure 9.2.3. Diesel cycle with super-charging.

Diesel Cycle 25

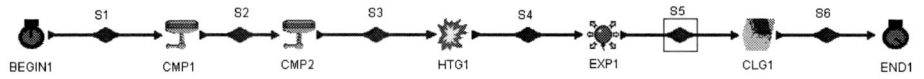

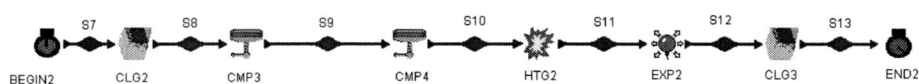

Figure 9.2.4. Diesel cycle with super-charging and pre-cooling.

EXAMPLE 9.2.3

Find the pressure and temperature of each state of an ideal Diesel cycle with a compression ratio of 15 and a cut-off ratio of 2. The cylinder volume before compression is 0.16 ft^3. The atmosphere conditions are 14.7 psia and 70°F. Also determine the mass of air in the cylinder, heat supplied, net work produced, MEP, and cycle efficiency.

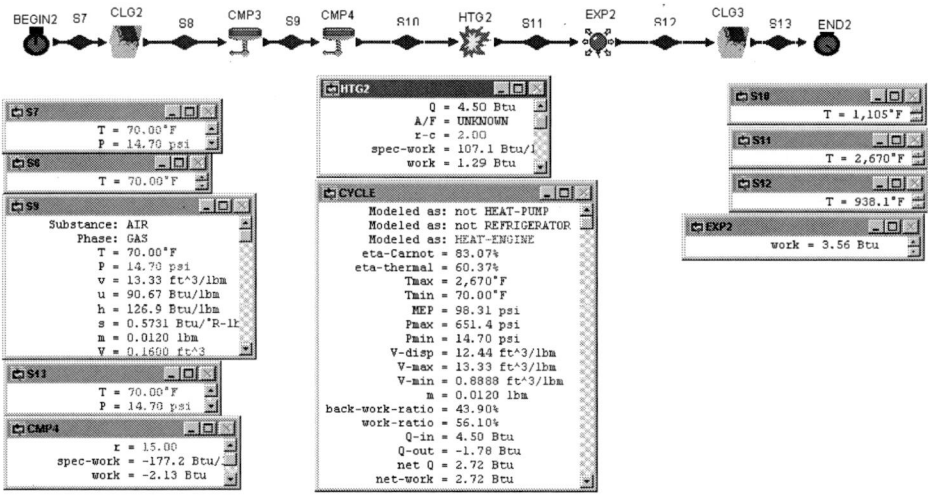

Figure E9.2.3. Diesel cycle without pre-cooler and without turbo-charger.

To solve this problem, we build the cycle as shown in Figure 9.2.4. Then (A) Assume isobaric for the pre-cooling process 7-8, isentropic for compression process 8-9, isentropic for compression process 9-10, isobaric for the heating process 10-11, isentropic for expansion process 11-12, and isochoric for the cooling process 12-13; (B) input p_7=14.7 psia, T_7=70°F; p_{13}=14.7 psia, T_{13}=70°F; p_8=14.7 psia, T_8=70°F; p_9=14.7 psia, V_9=0.16 ft^3 (no turbo-charger and no pre-cooler); compression ratio=15, and cut-off ratio=2; and (C) display results. The results are: T_8=70°F, T_9=96.87°F, T_{10}=1105°F, T_{11}=2670°F, T_{12}=938.1°F, Q_{in}=4.5 Btu, W_{net}=2.72 Btu, MEP=98.31 psia, η=60.37%, and m=0.012 lbm.

EXAMPLE 9.2.4

Find the pressure and temperature of each state of an ideal Diesel cycle with a compression ratio of 15 and a cut-off ratio of 2, and a super-charger which compresses fresh air to 20 psia before it enters the cylinder of the engine. The cylinder volume before compression is 0.16 ft^3. The atmosphere conditions are 14.7 psia and 70°F. Also determine the mass of air in the cylinder, heat supplied, net work produced, MEP, and cycle efficiency.

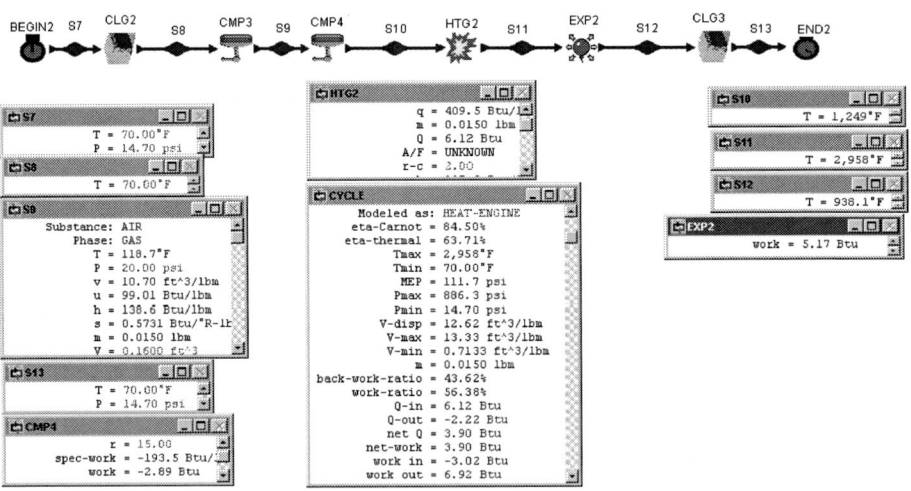

Figure E9.2.4. Diesel cycle with turbo-charger.

To solve this problem, we build the cycle as shown in Figure 9.2.4. Then (A) Assume isobaric for the pre-cooling process 7-8, isentropic for compression process 8-9, isentropic for compression process 9-10, isobaric for the heating process 10-11, isentropic for expansion process 11-12, and isochoric for the

Diesel Cycle

cooling process 12-13; (B) input p_7=14.7 psia, T_7=70°F; p_{13}=14.7 psia, T_{13}=70°F; T_8=70°F; p_9=20 psia, V_9=0.16 ft^3 (with turbo-charger and no pre-cooler); compression ratio=15, and cut-off ratio=2; and (C) display results. The results are: T_8=70°F, T_9=96.87°F, T_{10}=1249°F, T_{11}=2958°F, T_{12}=938.1°F, Q_{in}=6.12 Btu, W_{net}=3.90 Btu, MEP=106.5 psia, η=64.51%, and m=0.015 lbm.

EXAMPLE 9.2.5.

Find the pressure and temperature of each state of an ideal Diesel cycle with a compression ratio of 15 and a cut-off ratio of 2. A pre-cooler which cools the atmospheric air from 70°F to 50°F, and a super-charger which compresses fresh air to 20 psia before it enters the cylinder of the engine are added to the engine. The cylinder volume before compression is 0.16 ft^3. The atmosphere conditions are 14.7 psia and 70°F. Also determine the mass of air in the cylinder, heat supplied, net work produced, MEP, and cycle efficiency.

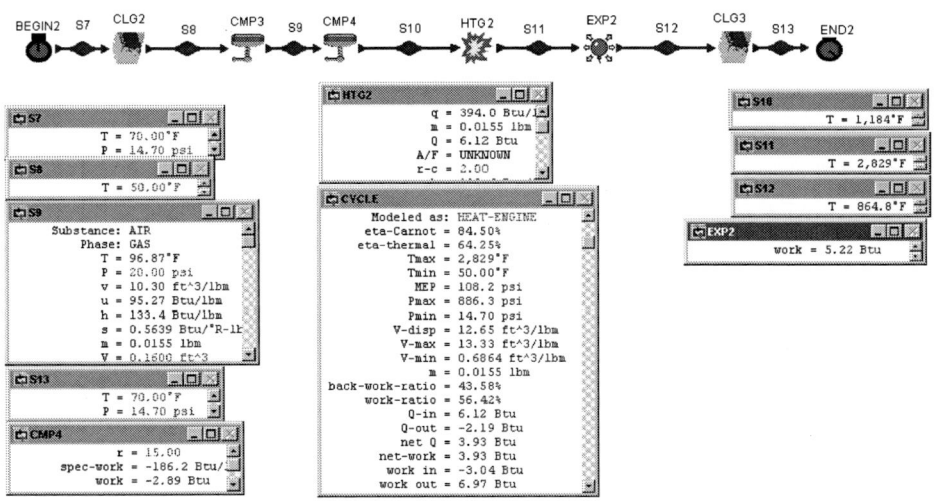

Figure E9.2.5. Diesel cycle with pre-cooler and turbo-charger.

To solve this problem, we build the cycle as shown in Figure 9.2.5. Then (A) Assume isobaric for the pre-cooling process 7-8, isentropic for compression process 8-9, isentropic for compression process 9-10, isobaric for the heating process 10-11, isentropic for expansion process 11-12, and isochoric for the cooling process 12-13; (B) input p_7=14.7 psia, T_7=70°F; p_8=14.7 psia, p_{13}=14.7

psia, $T_{13}=70°F$; $T_8=50°F$; $p_9=20$ psia, $V_9=0.16$ ft^3 (with turbo-charger and pre-cooler); compression ratio=15, and cut-off ratio=2; and (C) display results. The results are: $T_8=50°F$, $p_8=14.7$ psia, $T_9=96.87°F$, $p_9=20$ psia, $T_{10}=1184°F$, $p_{10}=886.3$ psia, $T_{11}=2829°F$, $p_{11}=886.3$ psia, $T_{12}=864.8°F$, $p_{12}=36.76$ psia; $Q_{78}=-0.0745$ Btu, $W_{89}=-0.1247$ Btu, $W_{910}=-2.89$ Btu, $W_{1011}=1.75$ Btu, $Q_{1011}=6.12$ Btu, $W_{1112}=5.2$ Btu, $W_{net}=3.93$ Btu, MEP=108.2 psia, $\eta=64.25\%$, and m=0.0155 lbm.

HOMEWORK 9.2. DIESEL CYCLE ANALYSIS AND OPTIMIZATION.

1. What is the difference between the compression ratio and cut-off ratio?
2. What is the difference between the Otto and Diesel engine?
3. How is the fuel introduced into the Diesel engine?
4. Does the Diesel engine have sparkling plugs? If yes, for what reason?
5. Does the Diesel engine have engine knock or detonation problem? Why?
6. Is the Otto cycle more efficient than a Diesel cycle with the same compression ratio?
7. How is it possible for a Diesel engine to operate at efficiencies greater than the efficiency of an Otto cycle?
8. Why is the Diesel engine usually used for big trucks and the Otto engine usually used for compact cars?
9. Can the Diesel engine afford to have a large compression ratio? Why?
10. How does the modern Diesel engine achieve higher power output without the use of higher compression ratio?
11. Suppose a large amount of power is required. Which engine would you choose between Otto and Diesel? Why?
12. In an ideal Diesel cycle, indicate whether the following statements are true or false:
    All the processes are internally reversible.
    Cycle efficiency increases with the maximum temperature.
    Cycle efficiency depends on the compression ratio only.
13. Sketch T-s and p-v diagrams for the Diesel cycle.
14. For a Diesel cycle, plot the cycle efficiency as a function of compression ratio from 4 to 30.
15. For a Diesel cycle, plot the MEP as a function of compression ratio from 4 to 30.
16. The compression ratio of an air-standard Diesel cycle is 15. At the beginning of the compression stroke, the pressure is 14.7 psia and the

Diesel Cycle

temperature is 80°F. The maximum temperature of the cycle is 4040°F. Find (A) the temperature at the end of the compression stroke, (B) the temperature at the beginning of the exhaust process, (C) the heat addition to the cycle, (D) the net work produced by the cycle, (E) the thermal efficiency, and (F) the MEP of the cycle.
ANSWER: (A) 1135°F, (B) 1847°F, (C) 1535 Btu/lbm, (D) 868.2 Btu/lbm, (E) 56.56%, (F) 167.8 psia.

17. An ideal Diesel cycle with a compression ratio of 17 and a cutoff ratio of 2 has an air temperature of 105°F and a pressure of 15 psia at the beginning of the isentropic compression process. Determine (A) the temperature and pressure of the air at the end of the isentropic compression process, (B) the temperature and pressure of the air at the end of the combustion process, and (C) the thermal efficiency of the cycle.
ANSWER: (A) 1294°F and 792.0 psia, (B) 3048°F and 792.0 psia, (C) 62.31%.

18. An ideal Diesel cycle with a compression ratio of 20 and a cutoff ratio of 2 has a temperature of 105°F and a pressure of 15 psia at the beginning of the compression process. Determine (A) the temperature and pressure of the gas at the end of the compression process, (B) the temperature and pressure of the gas at the end of the combustion process, (C) heat added to the engine in the combustion chamber, (D) heat removed from the engine to the environment, and (E) thermal cycle efficiency.
ANSWER: (A) 1412°F and 994.3 psia, (B) 3283°F and 994.3 psia, (C) 448.5 Btu/lbm, (D) -158.4 Btu/lbm, (E) 64.48%.

19. The pressure and temperature at the start of compression in an air Diesel cycle are 101 kPa and 300 K. The compression ratio is 15. The amount of heat addition is 2000 kJ/kg of air. Determine (A) the maximum cycle pressure and maximum temperature of the cycle, and (B) the cycle thermal efficiency.
ANSWER: (A) 2879 K and 4476 kPa, (B) 54.79%.

20. An ideal Diesel engine receives air at 103.4 kPa and 27°C. Heat added to the air is 1016.6 kJ/kg, and the compression ratio of the engine is 13. Determine (A) the work added during the compression process, (B) the cut-off ratio, (C) the work done during the expansion process, (D) the heat removed from the air during the cooling process, (E) the MEP (mean effective pressure), and (F) the thermal efficiency of the cycle.
ANSWER: (A) -385 kJ/kg, (B) 2.21, (C) 673.5 kJ/kg, (D) -437.7 kJ/kg, (E) 735.6 kPa, (F) 56.94%.

21. An ideal Diesel engine receives air at 15 psia and 65°F. Heat added to the air is 160 Btu/lbm, and the compression ratio of the engine is 6. Determine (A) the work added during the compression process, (B) the cut-off ratio, (C) the work done during the expansion process, (D) the heat removed from the air during the cooling process, (E) the MEP (mean effective pressure), and (F) the thermal efficiency of the cycle.
ANSWER: (A) -94.10 Btu/lbm, (B) 1.62, (C) 167.2 Btu/lbm, (D) -86.87 Btu/lbm, (E) 36.64 psia, (F) 45.71%.

22. An ideal Diesel engine receives air at 100 kPa and 25°C. The maximum cycle temperature is 1460°C and the compression ratio of the engine is 16. Determine (A) the work done during the compression process, (B) the heat added to the air during the heating process, (C) the work done during the expansion process, (D) the heat removed from the air during the cooling process, and (E) the thermal efficiency of the cycle.
ANSWER: (A) -434.1 kJ/kg, (B) 832.2 kJ/kg, (C) 710.5 kJ/kg, (D) -318.0 kJ/kg, (E) 61.79%.

23. A Diesel engine receives air at 60°F and 14.7 psia. The compression ratio is 20. The amount of heat addition is 800 Btu/lbm. The mass of air contained in the cylinder is 0.02 lbm. Determine the maximum cycle temperature, heat added, heat removed, work added, work produced, net work produced, MEP and efficiency of the cycle.
ANSWER: T_{max}=4601°F, Q_{add}=16 Btu, Q_{remove}=-6.27 Btu, W_{add}=-4.12 Btu, $W_{expansion}$=13.85 Btu, Wnet=9.73 Btu, MEP=211.7 psia, and η=60.84%.

24. A Diesel engine receives air at 80°F and 14.7 psia. The compression ratio is 20. The amount of heat addition is 800 Btu/lbm. The mass of air contained in the cylinder is 0.02 lbm. Determine the maximum cycle temperature, heat added, heat removed, work added, work produced, net work produced, MEP and efficiency of the cycle.
ANSWER: T_{max}=4667°F, Q_{add}=16 Btu, Q_{remove}=-6.22 Btu, W_{add}=-4.28 Btu, $W_{expansion}$=14.05 Btu, Wnet=9.78 Btu, MEP=204.7 psia, and η=61.11%.

25. An ideal Diesel engine receives air at 15 psia, 70°F. The air volume is 7 ft^3 before compression. Heat added to the air is 200 Btu/lbm, and the compression ratio of the engine is 11. Determine (A) the work added during the compression process, (B) the maximum temperature of the cycle, (C) the work done during the expansion process, (D) the heat removed from the air during the cooling process, (E) the MEP (mean effective pressure), and (F) the thermal efficiency of the cycle.

Diesel Cycle

ANSWER: (A) -78.19 Btu/lbm, (B) 1757°F, (C) 139.8 Btu/lbm, (D) -45.53 Btu/lbm, (E) 52.32 psia, (F) 57.51%.

26. A Diesel cycle has a compression ratio of 18. Air intake conditions (prior to compression) are 72°F and 14.7 psia, and the highest temperature in the cycle is limited to 2500°F to avoid damaging the engine block. Calculate: (A) thermal efficiency, (B) net work, and (C) mean effective pressure. Compare engine efficiency to that of a Carnot cycle engine operating between the same temperatures.
ANSWER: (A) 64.35%, (B) 195.9 Btu/lbm, (C) 83.76 psia; 82.04%.

27. A Diesel engine is modeled with an ideal Diesel cycle with a compression ratio of 17. The following information is known:
Temperature prior to the compression process: 70°F.
Pressure prior to the compression process: 14.7 psia.
Heat added during the combustion process: 245 Btu/lbm.
(A) Determine the temperature and pressure at each process endpoint.
(B) Solve for the net cycle work (Btu/lbm).
(C) Solve for the thermal efficiency.
ANSWER: (A) 1185°F and 776.2 psia, 2208°F and 776.2 psia, 582.3°F and 28.92 psia, (B) 157.3 Btu/lbm, (C) 64.2%.

28. An ideal Diesel cycle with a compression ratio of 17 and a cutoff ratio of 2 has a temperature of 313 K and a pressure of 100 kPa at the beginning of the isentropic compression process. Use the cold air-standard assumptions, assume that k=1.4, determine (A) the temperature and pressure of the air at the end of the isentropic compression process and at the end of the combustion process, and (B) the thermal efficiency of the cycle.
ANSWER: (A) 972.1 K and 5280 kPa, 1944 K and 5280 kPa, (B) 62.31%.

29. Find the pressure and temperature of each state of an ideal Diesel cycle with a compression ratio of 15 and a cut-off ratio of 2. A pre-cooler which cools the atmospheric air from 80°F to 50°F, and a super-charger which compresses fresh air to 20 psia before it enters the cylinder of the engine are added to the engine. The cylinder volume before compression is 0.1 ft³. The atmosphere conditions are 14.7 psia and 80°F. Also determine the mass of air in the cylinder, heat supplied, net work produced, MEP, and cycle efficiency.
ANSWER: [50°F and 14.7 psia, 98.67°F and 20 psia, 1184°F and 886.3 psia, 2829°F and 886.3 psia, and 854.9°F and 35.81 psia], m=0.0097 lbm, Q=3.83 Btu, Wnet=2.47 Btu, MEP=106.5 psia, η=64.51%.

30. Find the pressure and temperature of each state of an ideal Diesel cycle with a compression ratio of 15 and a cut-off ratio of 2. A pre-cooler which cools the atmospheric air from 80°F to 50°F, and a super-charger which compresses fresh air to 25 psia before it enters the cylinder of the engine are added to the engine. The cylinder volume before compression is 0.1 ft^3. The atmosphere conditions are 14.7 psia and 80°F. Also determine the mass of air in the cylinder, heat supplied, net work produced, MEP, and cycle efficiency.
ANSWER: [50°F and 14.7 psia, 133.5°F and 25 psia, 1293°F and 1108 psia, 3045°F and 1108 psia, and 854.9°F and 35.81 psia], m=0.0114 lbm, Q=4.78 Btu, Wnet=3.19 Btu, MEP=116.4 psia, η=66.7%.

31. A diesel engine has a state before compression of 95 kPa, 290 K, a peak pressure of 6000 kPa, and a maximum temperature of 2400 K. Find the work input, work output, net work output, heat input, thermal efficiency, and mean effective pressure of the cycle.
ANSWER: -471.6 kJ/kg, 1373 kJ/kg, 901.8 kJ/kg, 1457 kJ/kg, 61.90%, 1087 kPa.

32. A diesel engine has a state before compression of 100 kPa, 290 K, a peak pressure of 5000 kPa, and a maximum temperature of 2400 K. Find the work input, work output, net work output, heat input, thermal efficiency, and mean effective pressure of the cycle.
ANSWER: -427.7 kJ/kg, 1316 kJ/kg, 888.5 kJ/kg, 1518 kJ/kg, 58.52%, 1138 kPa.

33. At the beginning of compression in a Diesel cycle T=300 K and p=200 kPa; after combustion is complete T=1500 K and p=7 Mpa. Find the work input, work output, net work output, heat input, thermal efficiency, and mean effective pressure of the cycle.
ANSWER: -390.6 kJ/kg, 782.7 kJ/kg, 392.1 kJ/kg, 657.3 kJ/kg, 59.65%, 985.8 kPa.

34. At the beginning of compression in a Diesel cycle T=300 K and p=100 kPa; after combustion is complete T=1500 K and p=5 Mpa. Find the work input, work output, net work output, heat input, thermal efficiency, and mean effective pressure of the cycle.
ANSWER: -442.5 kJ/kg, 814.1 kJ/kg, 371.6 kJ/kg, 584.6 kJ/kg, 63.57%, 460.3 kPa.

Chapter 3

ATKINSON CYCLE

A cycle called Atkinson cycle is similar to the Otto cycle except that the isochoric exhaust and intake process at the end of the Otto cycle power stroke is replaced by an isobaric process. The schematic diagram of the cycle is shown in Figure 9.3.1. The cycle is made of the following four processes:

- 1-2   isentropic compression
- 2-3   isochoric heat addition
- 3-4   isentropic expansion
- 4-1   isobaric heat removing

Applying the First law and Second law of thermodynamics of the closed system to each of the four processes of the cycle yields:

$W_{12} = \int p dV$ (9.3.1)

$Q_{12} - W_{12} = m(u_2 - u_1), Q_{12}=0$ (9.3.2)

$W_{23} = \int p dV = 0$ (9.3.3)

$Q_{23} - 0 = m(u_3 - u_2)$ (9.3.4)

$W_{34} = \int p dV$ (9.3.5)

$Q_{34} - W_{34} = m(u_4 - u_3), Q_{34} = 0$ (9.3.6)

$$W_{41} = \int pdV = p\, m(v_1 - v_4) \tag{9.3.7}$$

and

$$Q_{41} - W_{41} = m(u_1 - u_4) \tag{9.3.8}$$

The net work (W_{net}), which is also equal to net heat (Q_{net}), is

$$W_{net} = W_{12} + W_{34} + W_{41} = Q_{net} = Q_{23} + Q_{41} \tag{9.3.9}$$

The thermal efficiency of the cycle is

$$\eta = W_{net}/Q_{23} = Q_{net}/Q_{23} = 1 - Q_{41}/Q_{23} = 1 - (h_4 - h_1)/(u_3 - u_2) \tag{9.3.10}$$

This expression for thermal efficiency of the cycle can be simplified if air is assumed to be the working fluid with constant specific heats. Equation (9.3.10) is reduced to:

$$\eta = 1 - k(T_4 - T_1)/(T_3 - T_2) \tag{9.3.11}$$

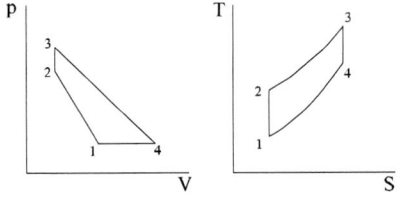

Figure 9.3.1. Atkinson cycle.

EXAMPLE 9.3.1

Find the pressure and temperature of each state of an ideal Atkinson cycle with a compression ratio of 8. The heat addition in the combustion chamber is 800 Btu/lbm. The atmospheric air is at 14.7 psia and 60°F. The cylinder contains 0.02 lbm of air. Determine the maximum temperature, maximum pressure, heat supplied, heat removed, work added during the compression processes, work produced during the expansion process, net work produced, MEP, and cycle efficiency. Draw the T-s diagram of the cycle.

To solve this problem, we build the cycle as shown in Figure E9.3.1. Then (A) Assume isentropic for the compression process 1-2 and the expansion process 3-4, isochoric for the heating process 2-3, and isobaric for the cooling process 4-1; (B) input p_1=14.7 psia, T_1=60°F, mdot=0.02 lbm; r=8 for the compression process 1-2, and q=800 Btu/lbm for the heating process 2-3. and (C) display results. The results are: T_{max}=5407°F, p_{max}=1328 psia, Q_{add}=16 Btu, Q_{remove}=-5.28 Btu, W_{comp}=-3.82 Btu, W_{expan}=14.54 Btu, W_{net}=10.72 Btu, MEP=74.00 psia, and η=67.02%.

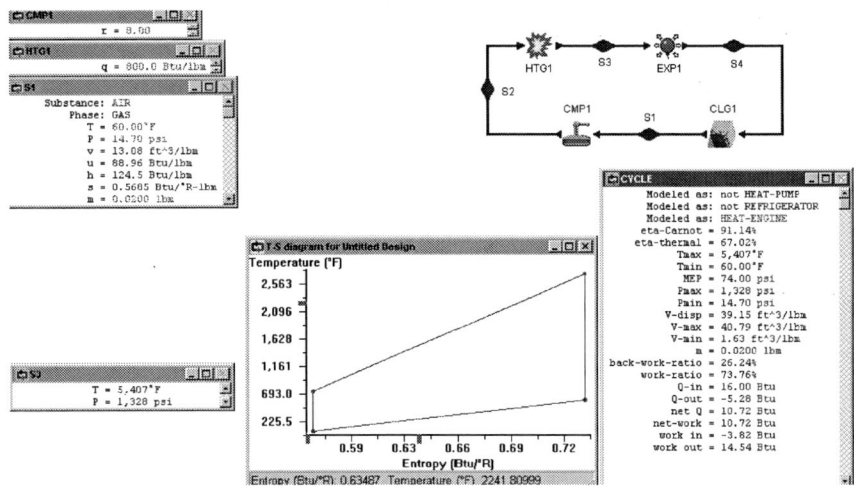

Figure E9.3.1. Atkinson cycle.

HOMEWORK 9.3. ATKINSON CYCLE

1. What are the four processes of the Atkinson cycle?
2. What is the difference between the Otto cycle and the Atkinson cycle?
3. Find the pressure and temperature of each state of an ideal Atkinson cycle with a compression ratio of 16. The heat addition in the combustion chamber is 800 Btu/lbm. The atmospheric air is at 14.7 psia and 60°F. The cylinder contains 0.02 lbm of air. Determine the maximum temperature, maximum pressure, heat supplied, heat removed, work added during the compression processes, work produced during the expansion process, net work produced, MEP, and cycle efficiency.

ANSWER: $T_{max}=5789°F$, $p_{max}=2828$ psia, $Q_{add}=16$ Btu, $Q_{remove}=-4.17$ Btu, $W_{comp}=-4.81$ Btu, $W_{expan}=16.63$ Btu, $W_{net}=11.83$ Btu, MEP=73.91 psia, and $\eta=73.91\%$.

4. Find the pressure and temperature of each state of an ideal Atkinson cycle with a compression ratio of 16. The heat addition in the combustion chamber is 800 Btu/lbm. The atmospheric air is at 101.4 kPa and 18°C. The cylinder contains 0.01 kg of air. Determine the maximum temperature, maximum pressure, heat supplied, heat removed, work added during the compression processes, work produced during the expansion process, net work produced, MEP, and cycle efficiency.
ANSWER: $T_{max}=3121°C$, $p_{max}=18904$ kPa, $Q_{add}=18$ kJ, $Q_{remove}=-4.72$ kJ, $W_{comp}=-5.59$ kJ, $W_{expan}=18.86$ kJ, $W_{net}=13.28$ kJ, MEP=631 kPa, and $\eta=73.75\%$.

5. Find the pressure and temperature of each state of an ideal Atkinson cycle with a compression ratio of 10. The heat addition in the combustion chamber is 800 Btu/lbm. The atmospheric air is at 101.4 kPa and 18°C. The cylinder contains 0.01 kg of air. Determine the maximum temperature, maximum pressure, heat supplied, heat removed, work added during the compression processes, work produced during the expansion process, net work produced, MEP, and cycle efficiency.
ANSWER: $T_{max}=2970°C$, $p_{max}=11289$ kPa, $Q_{add}=18$ kJ, $Q_{remove}=-5.54$ kJ, $W_{comp}=-4.74$ kJ, $W_{expan}=17.20$ kJ, $W_{net}=12.46$ kJ, MEP=540.7 kPa, and $\eta=69.21\%$.

6. An Atkinson cycle has a compression ratio of 10 and a top temperature in the cycle of 1400 K. The ambient temperature is 300 K.
   (A) What is the cycle efficiency?
   (B) What is the net work per unit of mass of this cycle?
   (C) Compare these values with those for an Otto cycle given the same conditions.

Chapter 4

DUAL CYCLE

Combustion in the Otto cycle is based on a constant volume process; in the Diesel cycle, it is based on a constant pressure process. But combustion in actual spark-ignition engine requires a finite amount of time if the process is to be completed. For this reason, combustion in Otto cycle does not actually occur under the constant volume condition. Similarly, in compression-ignition engines, combustion in Diesel cycle does not actually occur under the constant pressure condition, because of the rapid and uncontrolled combustion process.

The operation of the reciprocating internal combustion engines represents a compromise between the Otto and the Diesel cycle, and can be described as a Dual combustion cycle. Heat transfer to the system may be considered to occur first at constant volume and then at constant pressure. Such a cycle is called Dual cycle.

The Dual cycle as shown in Figure 9.4.1 is composed of the following five processes:

- 1-2   isentropic compression
- 2-3   constant volume heat addition
- 3-4   constant pressure heat addition
- 4-5   isentropic expansion
- 5-1   constant volume heat removing

Figure 9.4.2 shows the Dual cycle on p-v and T-s diagrams.

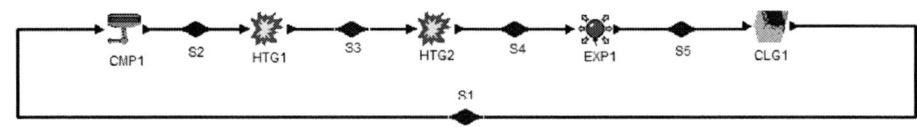

Figure 9.4.1. Dual cycle.

Applying the First law and Second law of thermodynamics of the closed system to each of the five processes of the cycle yields:

$$W_{12} = \int p dV \tag{9.4.1}$$

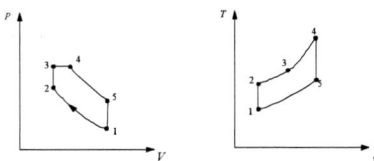

Figure 9.4.2. Dual cycle on p-v and T-s diagrams.

$$Q_{12} - W_{12} = m(u_2 - u_1), \quad Q_{12} = 0, \tag{9.4.2}$$

$$W_{23} = \int p dV = 0 \tag{9.4.3}$$

$$Q_{23} - 0 = m(u_3 - u_2) \tag{9.4.4}$$

$$W_{34} = \int p dV = m(p_4 v_4 - p_3 v_3) \tag{9.4.5}$$

$$Q_{34} = m(u_4 - u_3) + W_{34} = m(h_4 - h_3) \tag{9.4.6}$$

$$W_{45} = \int p dV \tag{9.4.7}$$

$$Q_{45} - W_{45} = m(u_5 - u_4), \quad Q_{45} = 0, \tag{9.4.8}$$

$$W_{51} = \int p dV = 0 \tag{9.4.9}$$

and

$$Q_{51} - W_{51} = m(u_1 - u_5) \tag{9.4.10}$$

The net work (W_{net}), which is also equal to net heat (Q_{net}), is

$$W_{net} = W_{12}+W_{34}+W_{45} = Q_{net} = Q_{23}+Q_{34}+Q_{51} \qquad (9.4.11)$$

The thermal efficiency of the cycle is

$$\eta = W_{net}/(Q_{23} + Q_{34}) = Q_{net}/(Q_{23} + Q_{34}) = 1 - Q_{51}/(Q_{23} + Q_{34}) \qquad (9.4.11)$$

This expression for thermal efficiency of an ideal Otto cycle can be simplified if air is assumed to be the working fluid with constant specific heats. Equation (9.4.11) is reduced to:

$$\eta = 1 - (T_5 - T_1)/[(T_3 - T_2) + k(T_4 - T_3)] \qquad (9.4.12)$$

EXAMPLE 9.4.1.

Pressure and temperature at the start of compression in a Dual cycle are 14.7 psia and 540°R. The compression ratio is 15. Heat addition at constant volume is 300 Btu/lbm of air, while heat addition at constant pressure is 500 Btu/lbm of air. The mass of air contained in the cylinder is 0.03 lbm. Determine (a) the maximum cycle pressure and maximum cycle temperature, (b) the efficiency and work output per kilogram of air, and (c) the MEP. Show the cycle on T-s diagram. Plot the sensitivity diagram of cycle efficiency vs compression ratio.

To solve this problem by CyclePad, we take the following steps:

1. Build
   (A) Take a compression device, two combustion chambers, an expander and a cooler from the closed system inventory shop and connect the five devices to form the Dual cycle.
   (B) Switch to analysis mode.
2. Analysis
   (A) Assume a process for each the five processes: (a) compression device as isentropic, (b) first combustion as isocbaric and second combustion as isobaric, (c) expander as isentropic, and (d) cooler as isochoric.
   (B) Input the given information: (a) working fluid is air, (b) the inlet pressure and temperature of the compression device are 14.7 psia and 540°R, (c) the compression ratio of the compression device is 15, (d)

the heat addition is 300 Btu/lbm in the isocbaric combustion chamber, (e) the heat addition is 500 Btu/lbm in the isobaric combustion chamber, and (f)The mass of air contained in the cylinder is 0.03 lbm.
3. Display results
   (A) Display the T-s diagram and cycle properties results. The cycle is a heat engine. The answers are T_{max}=5434°R, p_{max}=1367 psia, η=63.78%, MEP=217.3 psia and Wnet=15.31 Btu, and
   (B) Display the sensitivity diagram of cycle efficiency vs compression ratio.

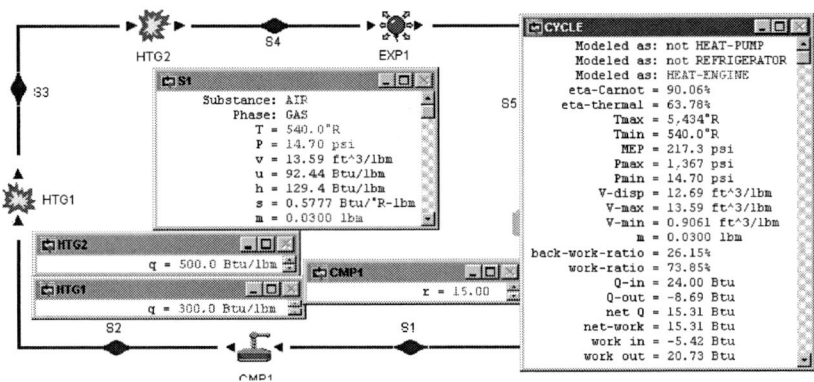

Figure E9.4.1a. Dual cycle.

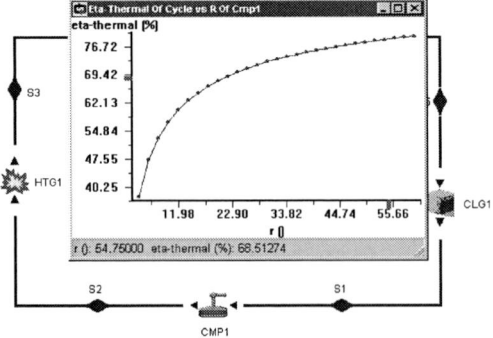

Figure E9.4.1b. Dual cycle sensitivity analysis.

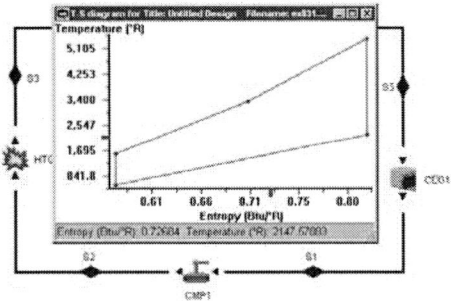

Figure E9.4.1c. Dual cycle T-s diagram.

HOMEWORK 9.4. DUAL CYCLE

1. What five processes make up the Dual cycle?
2. The combustion process in internal combustion engines as an isobaric or isometric heat addition process is over simplistic and not realistic. A real cycle p-v diagram of the Otto or Diesel cycle looks like a curve (combination of isobaric and isometric) rather than a linear line. Are the combustion processes in the Dual cycle more realistic?
3. Can we consider the Otto or Diesel cycle to be special cases of the Dual cycle?
4. Sketch T-s and p-v diagrams for the Dual cycle.
5. Show how the Dual cycle is a compromise between the Otto and Diesel cycles.
6. For a Dual cycle, plot the cycle efficiency as a function of compression ratio from 4 to 16.
7. For a Dual cycle, plot the MEP as a function of compression ratio from 4 to 16.
8. Pressure and temperature at the start of compression in a Dual cycle are 101 kPa and 15°C. The compression ratio is 8. Heat addition at constant volume is 100 kJ/kg of air, while the maximum temperature of the cycle is limited to 2000°C. The mass of air contained in the cylinder is 0.01 kg. Determine (a) the maximum cycle pressure, the MEP, Heat added, heat removed, compression work added, expansion work produced, net work produced and efficiency of the cycle.
ANSWER: p_{max}=2248 kPa, MEP=988.1 kPa, Q_{add}=15.77 kJ, Q_{remove}=-8.7 kJ, W_{comp}=-2.68 kJ, $W_{expansion}$=9.75 kJ, Wnet=7.07 kJ, and η=44.85%.

9. Pressure and temperature at the start of compression in a Dual cycle are 101 kPa and 15°C. The compression ratio is 12. Heat addition at constant volume is 100 kJ/kg of air, while the maximum temperature of the cycle is limited to 2000°C. The mass of air contained in the cylinder is 0.01 kg. Determine (a) the maximum cycle pressure, the MEP, Heat added, heat removed, compression work added, expansion work produced, net work produced and efficiency of the cycle.
ANSWER: p_{max}=3862 kPa, MEP=1067 kPa, Q_{add}=14.60 kJ, Q_{remove}=-6.60 kJ, W_{comp}=-kJ, $W_{expansion}$=11.51 kJ, Wnet=8.00 kJ, and η=57.48%.

10. Pressure and temperature at the start of compression in a Dual cycle are 101 kPa and 15°C. The compression ratio is 12. Heat addition at constant volume is 100 kJ/kg of air, while the maximum temperature of the cycle is limited to 2200°C. The mass of air contained in the cylinder is 0.01 kg. Determine (a) the maximum cycle pressure, the MEP, Heat added, heat removed, compression work added, expansion work produced, net work produced and efficiency of the cycle.
ANSWER: p_{max}=3862 kPa, MEP=1189 kPa, Q_{add}=16.60 kJ, Q_{remove}=-7.69 kJ, W_{comp}=-3.51 kJ, W_{comp}=12.43 kJ, Wnet=8.92 kJ, and η=53.71%.

Chapter 5

LENOIR CYCLE

The first commercially successful internal combustion engine was made by the French engineer Lenoir in 1860. He converted a reciprocating steam engine to admit a mixture of air and methane during the first half of the piston's outward suction stroke, at which point it was ignited with an electric spark and resulting combustion pressure acted on the piston for the remainder of the outward expansion stroke. The following inward stroke of the piston was used to expel the exhaust gases, and then the cycle began over again. The *Lenoir cycle* as shown in Figure 9.5.1 is composed of the following three effective processes:

    1-2   isochoric combustion process
    2-3   isentropic power expansion process
    3-1   isobaric exhaust process

The p-v and T-s diagrams of the cycle is shown in Figure 9.5.2.

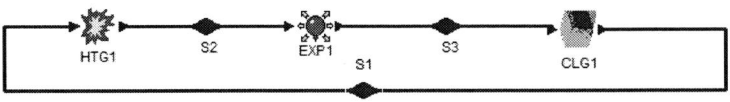

Figure 9.5.1. Lenoir cycle.

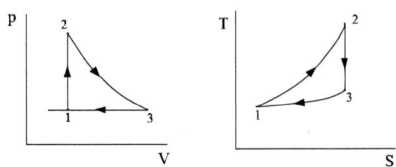

Figure 9.5.2. Lenoir cycle p-v diagram and T-s diagram.

Applying the First law and Second law of thermodynamics of the closed system to each of the three processes of the cycle yields:

$$W_{12} = 0 \tag{9.5.1}$$

$$Q_{12} - 0 = m(u_2 - u_1) \tag{9.5.2}$$

$$Q_{23} = 0 \tag{9.5.3}$$

$$0 - W_{23} = m(u_3 - u_2) \tag{9.5.4}$$
$$W_{31} = \int p dV = m(p_1 v_1 - p_3 v_3) \tag{9.5.5}$$

and

$$Q_{31} = m(u_1 - u_3) + W_{31} = m(h_1 - h_3) \tag{9.5.6}$$

The net work (W_{net}), which is also equal to net heat (Q_{net}), is

$$W_{net} = W_{23} + W_{31} = Q_{net} = Q_{12} + Q_{31} \tag{9.5.7}$$

The thermal efficiency of the cycle is

$$\eta = W_{net}/Q_{12} = Q_{net}/Q_{12} = 1 + Q_{31}/Q_{12} \tag{9.5.8}$$

This expression for thermal efficiency of the cycle can be simplified if air is assumed to be the working fluid with constant specific heats. Equation (9.5.8) is reduced to:

$$\eta = 1 - (h_3 - h_1)/(u_2 - u_1) = 1 - k\, T_2\, (r_s - 1)/(T_2 - T_1) \tag{9.5.9}$$

where r_s is the isentropic volume compression ratio, $r_s=v_3/v_1$.

Because the air-fuel mixture was not compressed before ignition, the engine efficiency was very low and fuel consumption was very high. The fuel-air mixture was ignited by an electric spark inside the cylinder.

EXAMPLE 9.5.1.

The isochoric heating process of a Lenoir engine receives air at 15°C and 101 kPa. The air is heated to 2000°C. The mass of air contained in the cylinder is 0.01 kg. Determine the pressure at the end of the isochoric heating process, the temperature at the end of the isentropic expansion process, heat added, heat removed, work added, work produced, net work produced, and efficiency of the cycle. Draw the T-s diagram of the cycle.

To solve this problem by CyclePad, we take the following steps:

1. Build
   (A) Take a combustion chamber, an expander and a cooler from the closed system inventory shop and connect the three devices to form the Lenoir cycle.
   (B) Switch to analysis mode.
2. Analysis
   (A) Assume a process for each the three processes: (a) combustion as isochoric, (b) expander as isentropic, and (c) cooler as isobaric.
   (B) Input the given information: (a) working fluid is air, (b) the inlet pressure and temperature of the combustion device are 101 kPa and 15°C, (c) the temperature at the end of combustion device is 2000°C, and (d) the mass of air is 0.01 kg.
3. Display results
   (A) Display cycle properties results. The cycle is a heat engine. The answers are T_3=986.8 °C, p_2=796.8 kPa, Q_{add}=14.23 kJ, Q_{remove}=-9.75 kJ, W_{comp}=-2.79 kJ, W_{expan}=7.26 kJ, W_{net}=4.48 kJ, and η=31.46%; and
   (B) Display the T-s diagram.

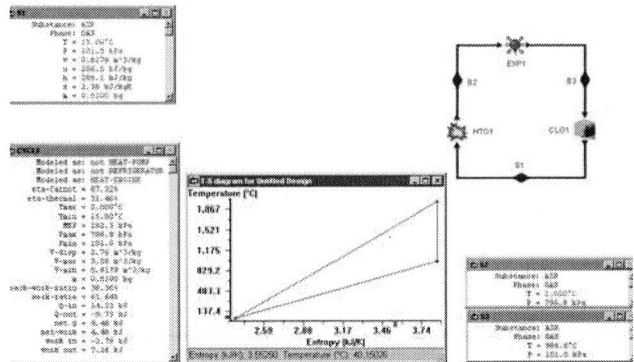

Figure E9.5.1. Lenoir cycle

HOMEWORK 9.5. LENOIR CYCLE

1. What are the five processes that make up the Lenoir cycle?
2. The isochoric heating process of a Lenoir engine receives air at 15°C and 101 kPa. The air is heated to 2200°C. The mass of air contained in the cylinder is 0.01 kg. Determine the pressure at the end of the isochoric heating process, the temperature at the end of the isentropic expansion process, heat added, heat removed, work added, work produced, net work produced, and efficiency of the cycle.
ANSWER: T_3=1065°C, p_2=866.9 kPa, Q_{add}=15.66 kJ, Q_{remove}=-10.54 kJ, W_{comp}=-3.01 kJ, W_{expan}=8.13 kJ, W_{net}=5.12 kJ, and η=32.72%.
3. The isochoric heating process of a Lenoir engine receives air at 60°F and 14.7 psia. The air is heated to 4000°F. The mass of air contained in the cylinder is 0.02 lbm. Determine the pressure at the end of the isochoric heating process, the temperature at the end of the isentropic expansion process, heat added, heat removed, work added, work produced, net work produced, and efficiency of the cycle.
ANSWER: T_3=1953°F, p_2=126.2 psia, Q_{add}=13.49 Btu, Q_{remove}=-9.08 Btu, W_{comp}=-2.59 Btu, W_{expan}=7.01 Btu, W_{net}=4.41 Btu, and η=32.72%.
4. The isochoric heating process of a Lenoir engine receives air at 80°F and 14.7 psia. The air is heated to 4500°F. The mass of air contained in the cylinder is 0.02 lbm. Determine the pressure at the end of the isochoric heating process, the temperature at the end of the isentropic expansion process, heat added, heat removed, work added, work produced, net work produced, and efficiency of the cycle.

ANSWER: T_3=2172°F, p_2=135.1 psia, Q_{add}=15.13 Btu, Q_{remove}=-10.03 Btu, W_{comp}=-2.86 Btu, W_{expan}=7.97 Btu, W_{net}=5.11 Btu, and η=33.74%.

Chapter 6

STIRLING CYCLE

The Stirling cycle is composed of the following four processes:

- 1-2   isothermal compression
- 2-3   constant volume heat addition
- 3-4   isothermal expansion
- 4-1   constant volume heat removing

Stirling-cycle engine is an external-combustion engine. Figure 9.6.1 shows the Stirling cycle on p-v and T-s diagrams.

During the isothermal compression process 1-2, heat is rejected to maintain a constant temperature T_L. During the isothermal expansion process 3-4, heat is added to maintain a constant temperature T_H. There are also heat interactions along the constant volume heat addition process 2-3 and the constant volume heat removing process 4-1. The quantities of heat in these two constant volume processes are equal but opposite in direction.

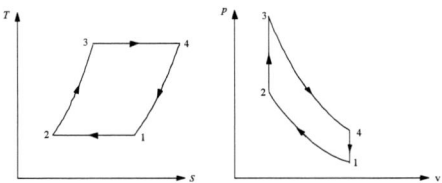

Figure 9.6.1. Stirling cycle on p-v and T-s diagrams.

The operation of the Stirling-cycle engine is shown in Figure 9.6.2. There are two pistons in the cylinder. One is a power piston (P), and the other is the displace piston (D). The purpose of the displace piston is to move the working fluid around from one space to another space through the regenerator. At state 1, the power piston is at BDC (bottom dead center), with the displacer at its TDC (top dead center). The power piston moves from its BDC to TDC to compress the working fluid during the compression process 1-2. From 1-2, the working fluid in the cylinder is in contact with the low temperature reservoir, so the temperature remains constant ($T_1=T_2$) and heat is removed. During the heating process 2-3, the displacer moves downward, pushing the working fluid through the regenerator where it picks up heat to reach T_3. During the expansion process 3-4, the working fluid in the cylinder is in contact with the high temperature reservoir, so the temperature remains constant ($T_3=T_4$) and heat is added. During the cooling process 4-1, the displacer moves upward, pushing the working fluid through the regenerator where it removes heat to reach T_1.

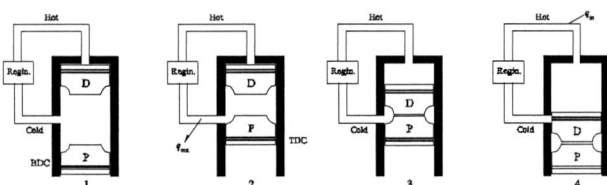

Figure 9.6.2. Stirling cycle operation.

Applying the First law and Second law of thermodynamics of the closed system to each of the four processes of the cycle yields:

$W_{12} = \int pdV$, $Q_{12} = \int TdS = T_1(S_2-S_1)$ (9.6.1)

$Q_{12} - W_{12} = m(u_2 - u_1) = 0$ (9.6.2)

$W_{23} = \int pdV = 0$ (9.6.3)

$Q_{23} - 0 = m(u_3 - u_2)$ (9.6.4)

$W_{34} = \int pdV$, $Q_{34} = \int TdS = T_3(S_4-S_3)$ (9.6.5)

$Q_{34} - W_{34} = m(u_4 - u_3) = 0$ (9.6.6)

$W_{41} = \int p dV = 0$ (9.6.7)

and

$Q_{41} - 0 = m(u_1 - u_4)$ (9.6.8)

The net work (W_{net}), which is also equal to net heat (Q_{net}), is

$W_{net} = W_{12} + W_{34} = Q_{net} = Q_{12} + Q_{23} + Q_{34} + Q_{41}$ (9.6.9)

The thermal efficiency of the cycle is

$\eta = W_{net} / (Q_{34} + Q_{23})$ (9.6.10)

EXAMPLE 9.6.1.

A Stirling cycle operates with 0.1 kg of hydrogen as a working fluid between 1000°C and 30°C. The highest pressure and the lowest pressure during the cycle are 3000 kPa and 500 kPa. Determine the heat and work added in each of the four processes, net work, and cycle efficiency.

To solve this problem by CyclePad, we take the following steps:

1. Build
   (A) Take a compression device, a combustion chamber, an expander and a cooler from the closed system inventory shop and connect the four devices to form the Stirling cycle.
   (B) Switch to analysis mode.
2. Analysis
   (A) Assume a process for each of the four processes: (a) compression device as isothermal, (b) combustion as isochoric, (c) expander as isothermal, and (d) cooler as isochoric.
   (B) Input the given information: (a) working fluid is helium, (b) the inlet pressure and temperature of the compression device are 500 kPa and 30°C and m=0.1 kg, (c) the inlet pressure and temperature of the expander are 3000 kPa and 1000°C.

3. Display results
   (A) Display the cycle properties results. The cycle is a heat engine. The answers are: $Q_{12}=W_{12}=-22.46$ kJ, $Q_{23}=300.7$ kJ, $Q_{34}=W_{34}=94.33$ kJ, $Q_{41}=-300.7$ kJ, $W_{net}=71.87$ kJ, $Q_{in}=395.0$ kJ, and $\eta=18.19\%$.

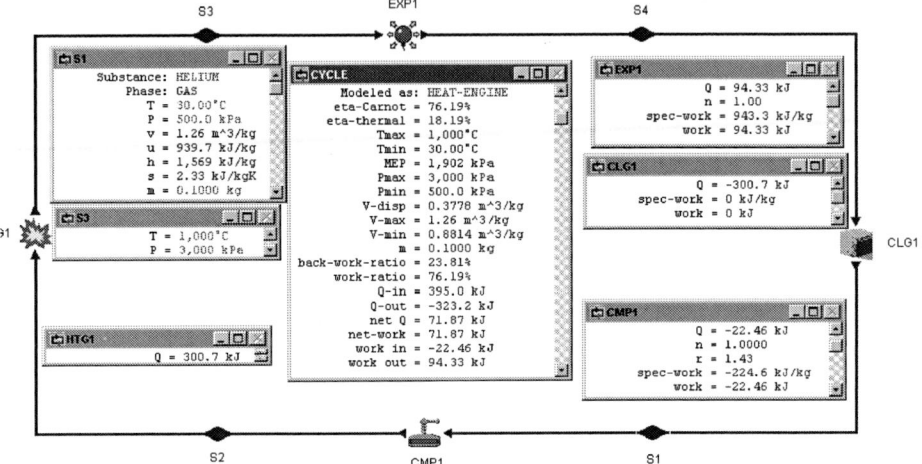

Figure E9.6.1. Stirling cycle.

The Stirling cycle is an attempt to achieve Carnot efficiency by the use of an ideal regenerator.

A device called regenerator can be used to absorb heat during process 4-1 (Q_{41}) and ideally delivering the same quantity of heat during process 2-3 (Q_{23}). These two quantities of heat are represented by the areas underneath of the process 4-1 and process 2-3 of the T-s diagram in Figure 9.6.1. Using the ideal regenerator, Q_{41} is not counted as a part of the heat input. The efficiency of the Stirling cycle can be reduced from Equation (9.6.10) to

$$\eta = W_{net}/Q_{12} = 1 - T_3/T_1 \qquad (9.6.11)$$

In this respect, the Stirling cycle has the same efficiency as the Carnot cycle.

The regenerative Stirling cycle is illustrated in Figure 9.6.3. In this figure, the combination of heater #1 and cooler #1 is equivalent to the regenerator. Heat removed from the cooler #1 is added to the heater #1. Since this energy transfer occurs within the cycle internally, the amount of heat added to the heater #1 from the cooler #1 is not a part of heat added to the cycle from its surrounding heat reservoirs. Therefore

$Q_{in} = Q_{12}$ (9.6.12)

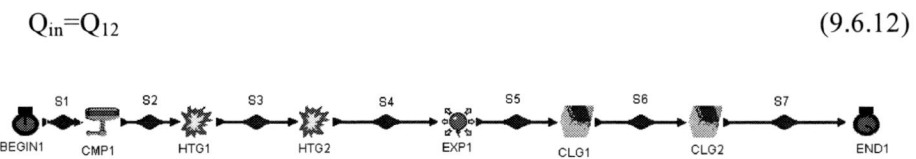

Figure 9.6.3. Regenerative Stirling cycle.

Practical attempts to follow the Stirling cycle present difficulties primarily due to the difficulty of achieving isothermal compression and isothermal expansion in a machine operating at a reasonable speed.

Example 9.6.2 illustrates the analysis of the regenerative Stirling cycle.

EXAMPLE 9.6.2.

A regenerative Stirling cycle operates with 0.1 kg of helium as a working fluid between 1000°C and 30°C. The highest pressure and the lowest pressure during the cycle are 3000 kPa and 500 kPa. The temperature at the exit of the regenerator (heater #1) and inlet to the heater #2 is 990°C and the temperature at the exit of the regenerator (cooler #1) and inlet to the cooler #2 is 40°C. Determine the heat and work added in each of the four processes, net work, and cycle efficiency.

To solve this problem by CyclePad, we take the following steps:

1. Build
   (A) Take a compression device, a heater (cold-side regenerator) combustion chamber, an expander and two coolers (cooler #1 is the hot-side regenerator) from the closed system inventory shop and connect the four devices to form the regenerative Stirling cycle as shown in Figure 9.6.2.
   (B) Switch to analysis mode.
2. Analysis
   (A) Assume a process for each of the six processes: (a) compression device as isothermal, (b) both heaters as isochoric, (c) expander as isothermal, and (d) both coolers as isochoric.
   (B) Input the given information: (a) working fluid is helium, (b) the inlet pressure and temperature of the compression device are 500 kPa and

30°C and m=0.1 kg, (c) the inlet pressure and temperature of the expander are 3000 kPa and 1000°C, (d) temperature at the exit of the regenerator (heater #1)= 990°C, (e) temperature at the exit of the regenerator (cooler #1)=40°C.
3. Display results
   (A) Display the cycle properties results. The cycle is a heat engine. The answers are: $Q_{12}=W_{12}=-22.46$ kJ, $Q_{23}=Q_{htr\#1}=-Q_{clr\#1}=Q_{regenerator}=297.6$ kJ, $Q_{34}=3.1$ kJ, $Q_{45}=W_{45}=94.33$ kJ, $Q_{56}=-3.1$ kJ, $W_{net}=71.87$ kJ, $Q_{in}=94.33+3.1=97.43$ kJ, and $\eta=71.87/97.43=73.77\%$.

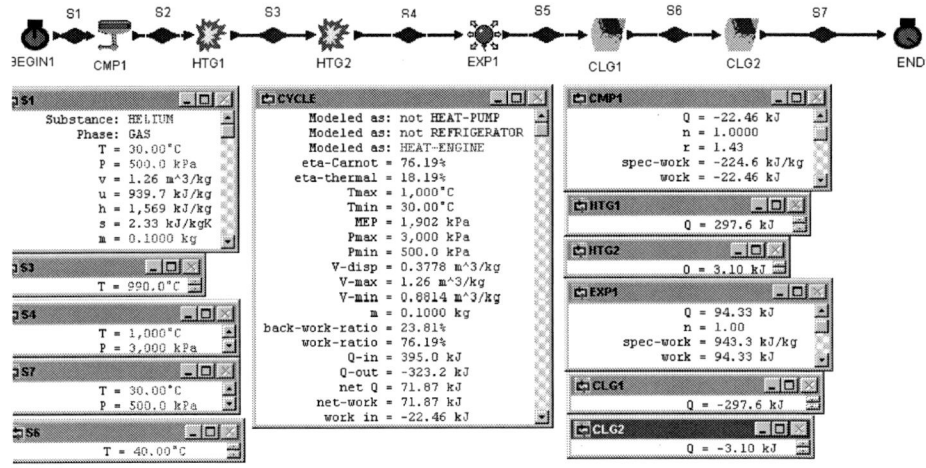

Figure E9.6.2. Regenerative Stirling cycle.

Comment: The regenerator used in this example is not ideal. Yet, the regenerator raises the cycle efficiency almost to the Carnot efficiency.

HOMEWORK 9.6. STIRLING CYCLE.

1. What are the four processes of the basic Stirling cycle?
2. The Stirling cycle uses a concept in its operation. Describe this concept and its principle.
3. What is a regenerator?
4. What is a regenerative Stirling cycle?

Stirling Cycle

5. What would be the cycle efficiency of the Stirling cycle with an ideal regenerator?
6. Sketch T-s and p-v diagrams for the Stirling cycle.
7. Theoretically, would there be any improvement in thermal cycle efficiency by the use of helium instead of air in a perfect Stirling cycle engine?
8. A Stirling cycle operates with 1 lbm of helium as a working fluid between 1800°R and 540°R. The highest pressure and the lowest pressure during the cycle are 450 psia and 75 psia. Determine the heat added, net work, and cycle efficiency.
   ANSWER: W_{net}=367.4 Btu, Q_{in}=1458 Btu, and η=25.20%.
9. A Stirling cycle operates with 0.1 kg of air as a working fluid between 1000°C and 30°C. The highest pressure and the lowest pressure during the cycle are 3000 kPa and 500 kPa. Determine the heat and work added in each of the four processes, net work, and cycle efficiency.
   ANSWER: Q_{12}=W_{12}=-3.10 kJ, Q_{23}=69.52 kJ, Q_{34}=W_{34}=13.02 kJ, Q_{41}=-69.52 kJ, W_{net}=9.92 kJ, Q_{in}=82.54 kJ, and η=12.02%.
10. A regenerative Stirling cycle operates with 0.1 kg of air as a working fluid between 1000°C and 30°C. The highest pressure and the lowest pressure during the cycle are 3000 kPa and 500 kPa. The temperature at the exit of the regenerator (heater #1) and inlet to the heater #2 is 990°C and the temperature at the exit of the regenerator (cooler #1) and inlet to the cooler #2 is 40°C. Determine the heat added, heat removed, work added, work removed, net work, MEP, and cycle efficiency.
    ANSWER: 82.54 kJ, -72.62 kJ, -3.10 kJ, 13.02 kJ, 9.92 kJ, 1902 kPa, 12.02%.
11. A regenerative Stirling cycle operates with 0.02 lbm of air as a working fluid between 1900°F and 80°F. The highest pressure and the lowest pressure during the cycle are 450 psia and 70 psia. The temperature at the exit of the regenerator (heater #1) and inlet to the heater #2 is 1800°F and the temperature at the exit of the regenerator (cooler #1) and inlet to the cooler #2 is 110°F. Determine the heat added, heat removed, work added, work removed, net work, MEP, and cycle efficiency.
    ANSWER: 7.48 Btu, -6.52 Btu, -2.849 Btu, 1.25 Btu, 0.9607 Btu, 284.5 psia, 12.85%.
12. A regenerative Stirling cycle operates with 0.02 lbm of carbon dioxide as a working fluid between 1900°F and 80°F. The highest pressure and the lowest pressure during the cycle are 450 psia and 70 psia. The temperature at the exit of the regenerator (heater #1) and inlet to the

heater #2 is 1800°F and the temperature at the exit of the regenerator (cooler #1) and inlet to the cooler #2 is 110°F. Determine the heat added, heat removed, work added, work removed, net work, MEP, and cycle efficiency.
ANSWER: 6.48 Btu, -5.85 Btu, -0.1877 Btu, 0.8208 Btu, 0.6331 Btu, 284.5 psia, 9.76%.

Chapter 7

MILLER CYCLE

Alternative to lowering the compression ratio and simultaneously the expansion ratio of an Otto or Diesel cycle, is to lower compression ratio only while the expansion ratio is kept as the original. *Miller* (reference: Miller, R.H., Supercharging and internal cooling cycle for high output, *Transaction of the American Society of Mechanical Engineers*, v69, pp453-457, 1947) proposed a cycle which has the following characteristics:

1. Effective compression stroke is shorter than expansion stroke
2. Increased charging pressure
3. Variable valve timing

Miller proposed the use of early intake valve closing to provide internal cooling before compression so as to reduce compression work. Miller further proposed increasing the boost pressure to compensate for the reduced inlet duration. By proper selection of boost pressure and variation of intake valve closing time, Miller showed that turbo-charged engines could maintain sea-level power while operating over varying altitudes.

A modified Otto cycle is known as the *Miller-Otto cycle* whose p-V and T-s diagrams are shown in Figure 9.7.1. A modified Diesel cycle is known as the *Miller-Diesel cycle* whose p-V and T-s diagrams are shown in Figure 9.7.2.

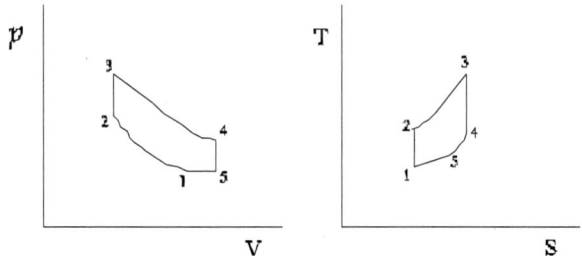

Figure 9.7.1. Miller-Otto cycle.

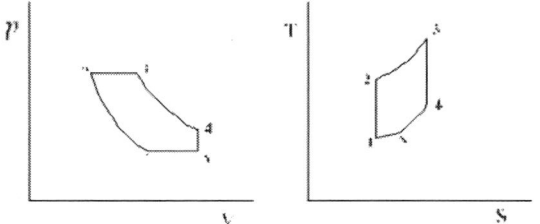

Figure 9.7.2. Miller-Diesel cycle.

A four stroke Miller-Otto cycle without supercharger and inter-cooler is shown in Figure 9.7.3. The intake valve is closed late at state 3.

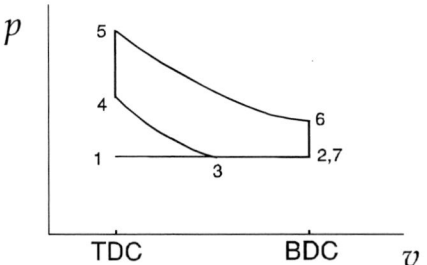

Figure 9.7.3. Miller-Otto cycle without supercharger and inter-cooler.

A four stroke Miller-Otto cycle with supercharger is shown in Figure 9.7.4. The intake valve is closed late at state 3.

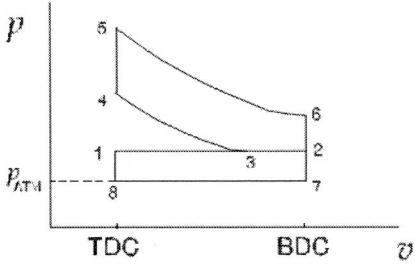

Figure 9.7.4. Miller-Otto cycle with supercharger.

Similarly, an extended expansion stroke is desirable in four-stroke spark ignition and Diesel engines from the viewpoint of providing an increase in thermal cycle efficiency and, for prescribed air and fuel flow rates, an increasing in engine output. Spark ignition engines modified to achieve extended expansion within the engine cylinder are termed Otto-Atkinson cycle (Reference: Ma, T.H., Recent advances in variable valve timing, *Automotive Engine Alternatives* Edited by R.L. Evans, Plenum Press, pp235-252, 1986). By analogy, Diesel engines modified to achieve extended expansion within the engine cylinder are termed Diesel-Atkinson cycle (Reference: Kentfield, J.A.C., Diesel engines with extended strokes, *SAE Transaction Journal of Engines*, v98, pp1816-1825, 1989).

Variable valve timing is being developed to improve the performance and reduce the pollution emissions from internal combustion heat engines for automobiles and trucks. A unique benefit for these engines is that changing the timing of the intake valves can be used to control the engine's compression ratio. These engines can be designed to have a conventionally high compression ratio for satisfactory cold starting characteristics and a reduced compression ratio for better cycle efficiency, exhaust gas emissions and noise characteristics.

EXAMPLE 9.7.1.

Determine the temperature at the end of the compression process, compression work, expansion work, and thermal efficiency of an ideal Otto cycle. The volume of the cylinder before and after compression are 3 liter and 0.3 liter. Heat added to the air in the combustion chamber is 800 kJ/kg. What is the mass of air in the cylinder? The atmosphere conditions are 101.3 kPa and 20°C.

To solve this problem, we build the Otto cycle. Then (A) Assume isentropic for compression process 1-2, isochoric for the heating process 2-3, isentropic for expansion process 3-4, and isochoric for the cooling process 4-5; (B) input p_1=101.3 kPa, T_1=20°C and V_1=3 L, V_2=0.3 L, heat added in the combustion chamber is 800 kJ/kg, p_5=101.3 kPa, and T_5=20°C; and (C) display results. The results are: T_2=463.2°C, W_{12}=-1.15 kJ, Q_{23}=2.89 kJ, W_{34}=2.89 kJ, W_{net}=1.74 kJ, η=60.19%, and m=3.62 g as shown in Figure E9.7.1.

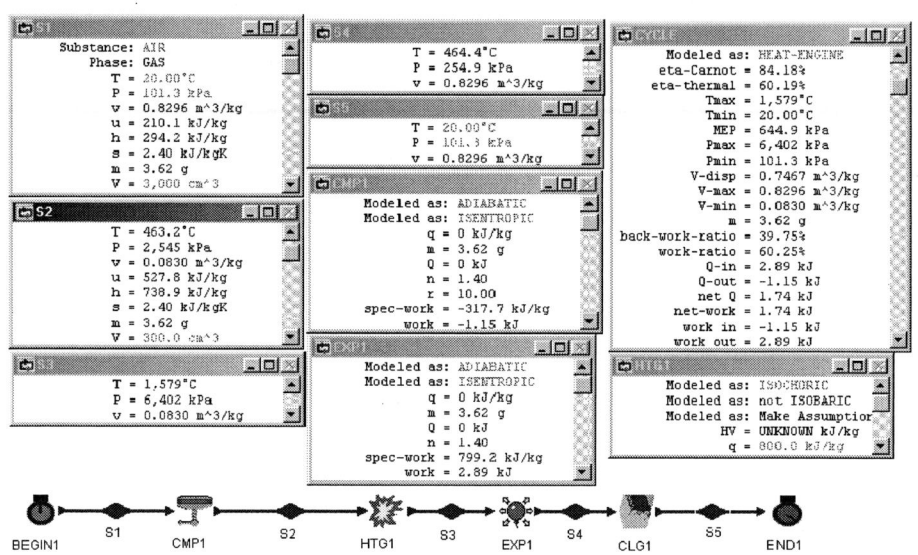

Figure E9.7.1. Otto cycle.

EXAMPLE 9.7.2.

Determine the temperature at the end of the compression process, compression work, expansion work, and thermal efficiency of an Otto-Miller cycle. The volume of the cylinder before and after compression are 3 liter and 0.3 liter. Heat added to the air in the combustion chamber is 800 kJ/kg. A supercharger and an inter-cooler are used. The supercharger pressure is 180 kPa and the temperature at the end of the inter-cooler is 20°C. The intake valve closes at 2.8 liter. The end temperature of the cooling process of the cycle is 20°C. What is the mass of air in the cylinder? The atmosphere conditions are 101.3 kPa and 20°C.

To solve this problem, we build the cycle as shown in Figure E9.7.2. Then (A) Assume isentropic for both compression processes, isochoric for the heating

Miller Cycle

process, isentropic for expansion process, and isochoric for both cooling processes; (B) input p_1=101.3 kPa and T_1=20°C, p_2=180 kPa, T_3=20°C and V_3=2.8 liter, V_4=0.3 liter, heat added in the combustion chamber is 800 kJ/kg, V_6=3 liter and T_7=20°C; and (C) display results. The results are: T_4=443.2°C, W_{comp}=-1.73 kJ, Q_{add}=4.07 kJ, W_{exp}=4.02 kJ, W_{net}=2.29 kJ, η=2.29/4.07=56.27%, and m=5.09 g as shown in Figure E9.7.2. Notice that if the supercharger is operated by the exhaust gas, then η=(4.02-1.54)/4.07=60.93%

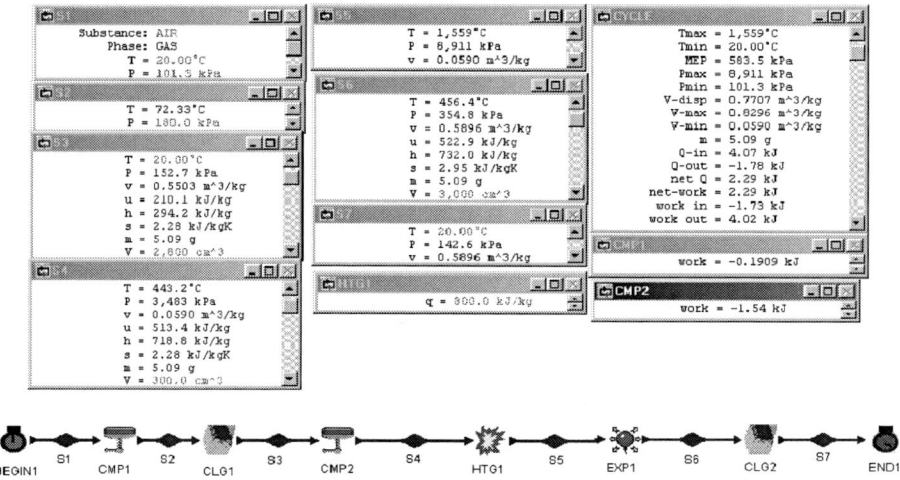

Figure E9.7.2. Miller-Otto cycle with supercharger and inter-cooler.

HOMEWORK 9.7. MILLER CYCLE

1. What is the idea of the Miller cycle?
2. What are the benefits of the Miller cycle?
3. Determine the temperature at the end of the compression process, compression work, expansion work, and thermal efficiency of an Otto-Miller cycle. The volume of the cylinder before and after compression are 3 liter and 0.3 liter. Heat added to the air in the combustion chamber is 800 kJ/kg. A supercharger and an inter-cooler are used. The supercharger pressure is 180 kPa and the temperature at the end of the inter-cooler is 20°C. The intake valve closes at 2.5 liter. The end temperature of the

cooling process of the cycle is 20°C. What is the mass of air in the cylinder? The atmosphere conditions are 101.3 kPa and 20°C.
ANSWER: T_4=411.4°C, W_{comp}=-1.44 kJ, Q_{add}=3.63 kJ, W_{exp}=3.53 kJ, W_{net}=2.08 kJ, η=57.3%, and m=4.54 g.

4. Determine the temperature at the end of the compression process, compression work, expansion work, and thermal efficiency of an Otto-Miller cycle. The volume of the cylinder before and after compression are 3 liter and 0.3 liter. Heat added to the air in the combustion chamber is 800 kJ/kg. A supercharger and an intercooler are used. The supercharger pressure is 200 kPa and the temperature at the end of the intercooler is 20°C. The intake valve closes at 2.0 liter. The end temperature of the cooling process of the cycle is 20°C. What is the mass of air in the cylinder? The atmosphere conditions are 101.3 kPa and 20°C.
ANSWER: T_4=353°C, W_{comp}=-1.11 kJ, Q_{add}=3.14 kJ, W_{exp}=2.95 kJ, W_{net}=1.83 kJ, η=58.28%, and m=3.92 g.

5. Determine the temperature at the end of the compression process, compression work, expansion work, and thermal efficiency of an Otto-Miller cycle. The volume of the cylinder before and after compression are 3 liter and 0.3 liter. Heat added to the air in the combustion chamber is 800 kJ/kg. A supercharger and an intercooler are used. The supercharger pressure is 180 kPa and the temperature at the end of the intercooler is 20°C. The intake valve closes at 2 liter. The end temperature of the cooling process of the cycle is 20°C. What is the mass of air in the cylinder? The atmosphere conditions are 101.3 kPa and 20°C.
ANSWER: T_4=411.4°C, W_{comp}=-1.44 kJ, Q_{add}=3.63 kJ, W_{exp}=3.53 kJ, W_{net}=2.08 kJ, η=57.3%, and m=4.54 g.

Chapter 8

WICKS CYCLE

The Carnot cycle is the ideal cycle only for the conditions of constant temperature hot and cold surrounding thermal reservoirs. However, the conditions of constant temperature hot and cold surrounding thermal reservoirs do not exist for fuel burning engines. For fuel burning engines, the combustion products are artificially created as a finite size hot reservoir that releases heat over the entire temperature range from its maximum to ambient temperature. The natural environment in terms of air or water bodies is the cold reservoir and can be considered as an infinite reservoir relative to the engine. Thus, an ideal fuel burning engine should operate reversibly between a finite size hot reservoir and an infinite size cold reservoir. Wicks (Reference: Wicks, F., The thermodynamic theory and design of an ideal fuel burning engine, Proceedings of the Intersociety Engineering Conference of Energy Conversion, v2, pp474-481, 1991) proposed a three-process ideal fuel burning engine consisting of an isothermal compression, an isochoric heat addition, and an adiabatic expansion process. The schematic Wicks cycle is shown in Figure 9.8.1. The p-v and T-s diagrams of the cycle is shown in Figure 9.8.2. and an example of the cycle is given in Example 9.8.1.

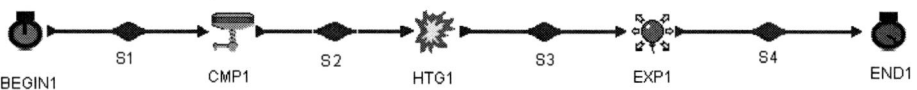

Figure 9.8.1. Wicks cycle.

64   Chih Wu

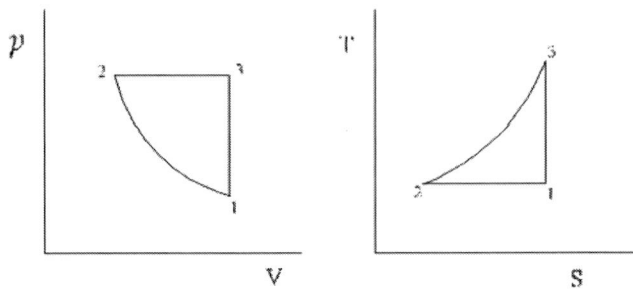

Figure 9.8.2. p-v and T-s diagrams of the Wicks cycle.

EXAMPLE 9.8.1.

Air is compressed from 14.7 psia and 500°R isothermally to 821.8 psia, heated isochorically to 2500°R, and then expanded isentropically to 14.7 psia in a Wick cycle. Determine the heat added, heat removed, work added, work produced, net work, and cycle efficiency.

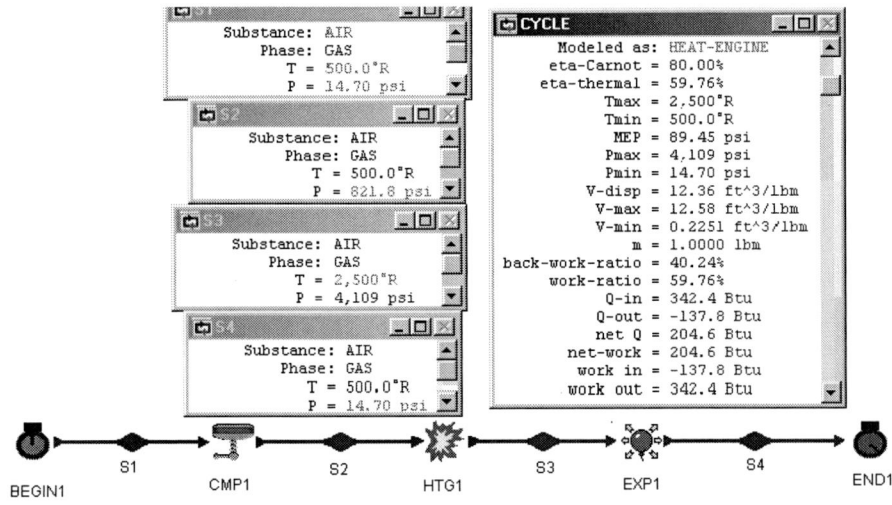

Figure E9.8.1. Wicks cycle.

To solve this problem by CyclePad, we take the following steps:

1. Build
   (A) Take a begin, an end, a compression device, a heater, and an expander from the closed system inventory shop and connect them to form the Wicks cycle as shown in Figure 9.8.1.
   (B) Switch to analysis mode.
2. Analysis
   (A) Assume a process for each of the three processes: (a) compression device as isothermal, (b) heater as isochoric, and (c) expander as isentropic.
   (B) Input the given information: (a) working fluid is air, (b) the begin pressure and temperature of the compression device are 14.7 psia and 500°R, and m=1 lbm, (c) the end temperature of the heater expander is 2500°R, and (d) the end pressure of the expander is 14.7 psia.
3. Display results
   (A) Display the cycle properties results. The cycle is a heat engine. The answers are: Q_{in}=342.4 Btu, Q_{out}=-137.8 Btu, W_{in}=-137.8 Btu, W_{out}=342.4 Btu, W_{net}=204.6 Btu, MEP=89.45 psia, and η=59.76%.

HOMEWORK 9.8. WICKS CYCLE

1. What are the processes of the Wicks cycle?
2. Air is compressed from 100 kPa and 20°C isothermally to 2000 kPa, heated isochorically to 1200°K, and then expanded isentropically to 100 kPa in a Wick cycle. Determine the heat added, heat removed, work added, work produced, net work, and cycle efficiency.

Chapter 9

RALLIS CYCLE

The Rallis cycle is defined by two isothermal processes at temperatures T_H and T_L separated by two regenerative processes which are part constant volume and part constant pressure in any given combination. The Stirling cycle is a special case of the Rallis cycle. Many other Rallis cycles can be defined which have no identifying names.

A conceptual arrangement of a Rallis heat engine is shown in Figure 9.9.1. The p-v and T-s diagrams for the cycle are shown in Figure 9.9.2. T_H is the heat source temperature and T_L is the heat sink temperature. The cycle is composed of the following six processes:

        1-2   isobaric cooling
        2-3   isothermal compression at T_L
        3-4   constant volume heat addition
        4-5   isobaric heating
        5-6   isothermal expansion at T_H
        6-1   constant volume heat removing

During the isothermal compression process 2-3, heat is rejected to maintain a constant temperature T_L. During the isothermal expansion process 5-6, heat is added to maintain a constant temperature T_H. There are heat interactions along the constant volume heat addition process 3-4 and the constant volume heat removing process 6-1; the quantities of heat in these two constant volume processes are equal but opposite in direction. There are also heat interactions along the constant pressure heat addition process 4-5 and the constant pressure heat removing process 1-2. The quantities of heat in these two constant pressure processes are equal but opposite in direction.

Applying the First law and Second law of thermodynamics of the closed system to each of the six processes of the cycle yields:

$$W_{12} = \int pdV = p_1(V_2-V_1) \tag{9.9.1}$$

$$Q_{12} - W_{12} = m(u_2 - u_1), \quad Q_{12} = -Q_{34} \tag{9.9.2}$$

$$Q_{23} = \int TdS = T_L(S_3-S_2) \tag{9.9.3}$$

$$Q_{23} - W_{23} = m(u_3 - u_2) = 0 \tag{9.9.4}$$

$$W_{34} = \int pdV = 0 \tag{9.9.5}$$

$$Q_{34} - 0 = m(u_4 - u_3), \quad Q_{34} = -Q_{12} \tag{9.9.6}$$

$$W_{45} = \int pdV = p_4(V_5-V_4) \tag{9.9.7}$$

$$Q_{45} - W_{45} = m(u_5 - u_4), \quad Q_{45} = -Q_{61}$$

$$Q_{56} = \int TdS = T_H(S_6-S_5) \tag{9.9.8}$$

$$Q_{56} - W_{56} = m(u_6 - u_5) = 0 \tag{9.9.9}$$

$$W_{61} = \int pdV = 0 \tag{9.9.10}$$

and

$$Q_{61} - 0 = m(u_1 - u_6), \quad Q_{61} = -Q_{45} \tag{9.9.11}$$

The net work (W_{net}), which is also equal to net heat (Q_{net}), is

$$W_{net} = W_{12} + W_{23} + W_{45} + W_{56} = Q_{net}$$

$$= Q_{12} + Q_{23} + Q_{34} + Q_{45} + Q_{56} + Q_{61}$$

$$= Q_{23} + Q_{56}. \tag{9.9.12}$$

The thermal efficiency of the cycle is

$$\eta = W_{net}/Q_{56} \qquad (9.9.13)$$

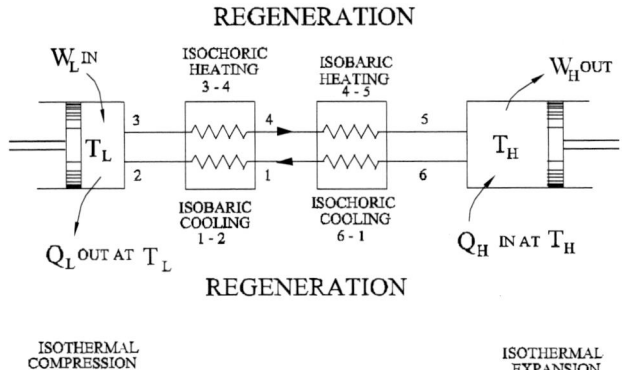

Figure 9.9.1. Rallis cycle.

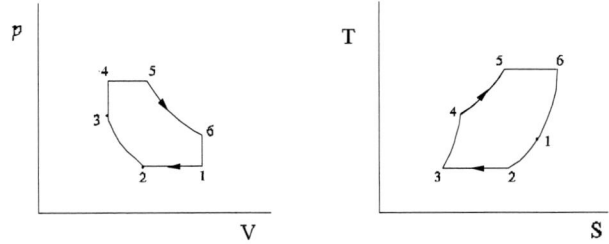

Figure 9.9.2. p-v and T-s diagram of Rallis cycle.

EXAMPLE 9.9.1.

A Rallis heat engine is shown in Figure E9.9.1a. Helium mass contained in the cylinder is 0.1 lbm. The six processes are:

        1-2   isobaric cooling
        2-3   isothermal compression at T_L
        3-4   constant volume heat addition
        4-5   isobaric heating

5-6   isothermal expansion at T_H
6-1   constant volume heat removing

The following information is given:

p_2=15 psia, T_2=60°F, q_{34}=60 Btu/lbm, q_{12}=-60 Btu/lbm, p_5=100 psia, and T_5=800°F.
Determine the pressure and temperature of each state of the cycle, work and heat of each process, work input, work output, net work output, heat added, heat removed, MEP and cycle efficiency. Draw the T-s diagram of the cycle.

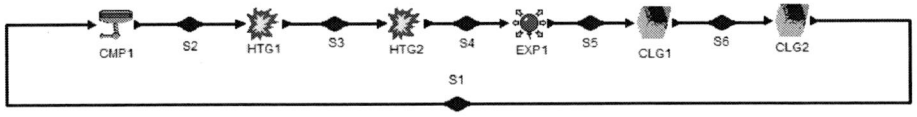

Figure E9.9.1a. Rallis heat engine.

To evaluate this example by CyclePad, we take the following steps:

1. Build
   (A) Take a compression device, two heaters, an expander and two coolers from the closed system inventory shop and connect the six devices to form the cycle as shown in Figure E9.9.1a.
   (B) Switch to analysis mode.
2. Analysis
   (A) Assume a process for each of the six processes: (a) compression device as isothermal, (b) one heater as isochoric and the other as isobaric, (c) expander as isothermal, and (d) one cooler as isochoric and the other as isobaric.
   (B) Input the given information: working fluid is air, m=0.1 lbm, p_2=15 psia, T_2=60°F, q_{34}=60 Btu/lbm, q_{12}=-60 Btu/lbm, p_5=100 psia, and T_5=800°F as shown in Figure E9.9.1b.
3. Display results
   (A) Display the cycle properties results. The cycle is a heat engine. The results are: p_1=15 psia, T_1=108.5°F, p_2=15 psia, T_2=60°F, p_3=45.11 psia, T_3=60°F, p_4=45.11 psia, T_4=108.5°F, p_5=100 psia, T_5=800°F, p_6=33.25 psia, T_6=800°F; q_{12}=-60 Btu/lbm, w_{12}=-24.07 Btu/lbm, q_{23}=-283.8 Btu/lbm, w_{23}=-283.8 Btu/lbm, q_{34}=60 Btu/lbm, w_{34}=2.41 Btu/lbm, q_{45}=512 Btu/lbm, w_{45}=0 Btu/lbm, q_{56}=688 Btu/lbm,

Rallis Cycle

w_{56}=688 Btu/lbm, q_{61}=-512 Btu/lbm, w_{61}=0 Btu/lbm; Q_{in}=126 Btu, Q_{out}=-85.58 Btu, Q_{net}=40.42 Btu, W_{in}=-30.79 Btu, W_{out}=71.21 Btu, W_{net}=40.42 Btu, MEP=30.91 psia, and η=32.08% as shown in Figure E9.9.1c. The T-s diagram of the cycle is shown in Figure E9.9.1d.

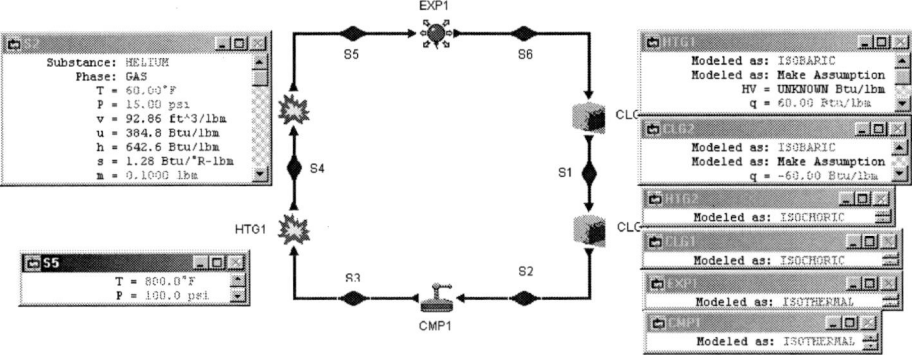

Figure E9.9.1b. Rallis heat engine input.

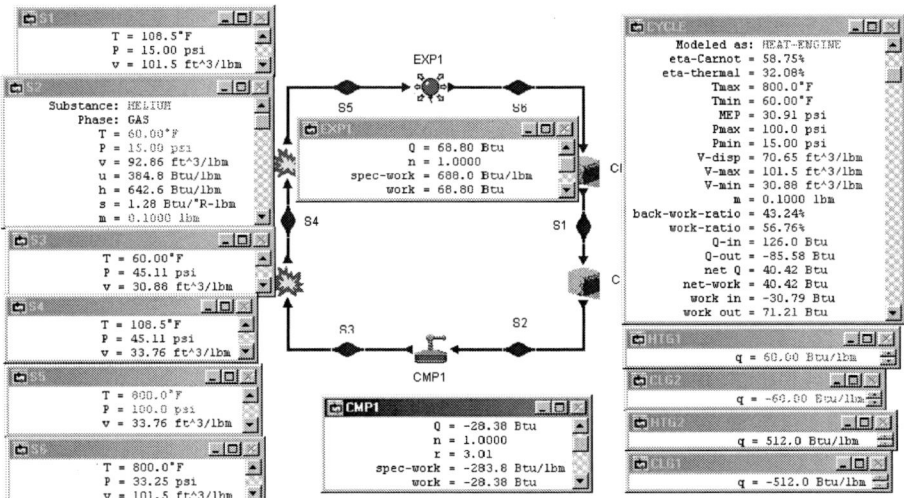

Figure E9.9.1c. Rallis cycle output results.

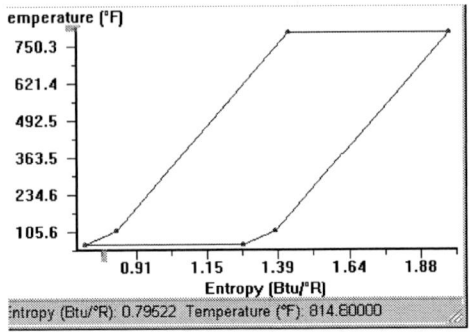

Figure E9.9.1d. Rallis cycle T-s diagram.

HOMEWORK 9.9. RALLIS CYCLE

1. What is the Rallis cycle?
2. How many regenerating processes are there in a Rallis cycle?
3. A Rallis heat engine is shown in Figure E9.9.1a. Air mass contained in the cylinder is 0.1 lbm. The six processes are:
   1-2   isobaric cooling
   2-3   isothermal compression at T_L
   3-4   constant volume heat addition
   4-5   isobaric heating
   5-6   isothermal expansion at T_H
   6-1   constant volume heat removing
   The following information is given:
   p_2=15 psia, T_2=60°F, q_{34}=60 Btu/lbm, q_{12}=-60 Btu/lbm, p_5=100 psia, and T_5=800°F.
   Determine the pressure and temperature of each state of the cycle, work and heat of each process, work input, work output, net work output, heat added, heat removed, MEP and cycle efficiency. Draw the T-s diagram of the cycle.
   ANSWER: [(15 psia, 310.4°F), (15 psia, 60°F), (59.72 psia, 60°F), (100 psia, 410.5°F), (100 psia, 800°F), (24.54 psia, 800°F)], [(-1.71 Btu, -6 Btu), (-4.92 Btu, -4.92 Btu), (0 Btu, 6 Btu), (2.67 Btu, 9.34 Btu), (12.12 Btu, 12.12 Btu), (0 Btu, -8.38 Btu)], -6.63 Btu, 14.79 Btu, 8.15 Btu, 27.45 Btu, -19.3 Btu, 27.93 psia, 29.70%.

Rallis Cycle

4. A Rallis heat engine is shown in Figure E9.9.1a. Carbon dioxide mass contained in the cylinder is 0.1 lbm. The six processes are:
   1-2   isobaric cooling
   2-3   isothermal compression at T_L
   3-4   constant volume heat addition
   4-5   isobaric heating
   5-6   isothermal expansion at T_H
   6-1   constant volume heat removing
   The following information is given:
   p_2=15 psia, T_2=60°F, q_{34}=60 Btu/lbm, q_{12}=-60 Btu/lbm, p_5=100 psia, and T_5=800°F.
   Determine the work input, work output, net work output, heat added, heat removed, MEP and cycle efficiency. Draw the T-s diagram of the cycle.
   ANSWER: -4.50 Btu, 9.93 Btu, 5.44 Btu, 21.45 Btu, -16.01 Btu, 26.47 psia, 58.75%.

5. A Rallis heat engine is shown in Figure E9.9.1a. Air mass contained in the cylinder is 0.1 lbm. The six processes are:
   1-2   constant volume heat removing
   2-3   isothermal compression at T_L
   3-4   isobaric heating
   4-5   constant volume heat addition
   5-6   isothermal expansion at T_H
   6-1   isobaric cooling
   The following information is given:
   p_2=15 psia, T_2=60°F, q_{34}=60 Btu/lbm, q_{12}=-60 Btu/lbm, p_5=100 psia, and T_5=800°F.
   Determine the work input, work output, net work output, heat added, heat removed, MEP and cycle efficiency. Draw the T-s diagram of the cycle.
   ANSWER: -6.63 Btu, 14.79 Btu, 8.15 Btu, 27.45 Btu, -19.3 Btu, 27.93 psia, 29.70%.

6. A Rallis heat engine is shown in Figure E9.9.1a. Air mass contained in the cylinder is 0.01 kg. The six processes are:
   1-2   constant volume heat removing
   2-3   isothermal compression at T_L
   3-4   isobaric heating
   4-5   constant volume heat addition
   5-6   isothermal expansion at T_H
   6-1   isobaric cooling
   The following information is given:

p_2=100 kPa, T_2=15°C, q_{34}=140 kJ/kg, q_{12}=-140 kJ/kg, p_5=700 kPa, and T_5=430°C.

Determine the work input, work output, net work output, heat added, heat removed, MEP and cycle efficiency. Draw the T-s diagram of the cycle.
ANSWER: -1.58 kJ, 3.55 kJ, 1.97 kJ, 6.52 kJ, -455 kJ, 191.6 kPa, 30.20%.

Chapter 10

DESIGN EXAMPLES

Although the Carnot cycle is useful in determining the ideal behavior of ideal heat engine, it is not a practical cycle to use in the design of heat engines. There are different reasons for developing cycles other than the Carnot cycle. These reasons includes the characteristics of the energy source available, working fluid chosen for the cycle, material limitations in the hardware and other practical consideration.

CyclePad is a powerful tool for cycle design and analysis. Due to its capabilities, the software allows users to view the cycle effects of varying design input parameters at once. The following examples illustrate the design of several closed-system gas power cycles.

EXAMPLE 9.10.1.

A six-process internal combustion engine as shown in Figure E9.10.1a is proposed by a junior engineer. Air mass contained in the cylinder is 0.01 kg. The six processes are:

Process 1-2 isentropic compression
Process 2-3 isochoric heating
Process 3-4 isobaric heating
Process 4-5 isentropic expansion
Process 5-6 isochoric cooling
Process 6-1 isobaric cooling

The following information is given:

p_1=100 kPa, T_1=20°C, V_1=10V_2, q_{23}=600 kJ/kg, q_{34}=400 kJ/kg, and T_5=400°C.

Determine the pressure and temperature of each state of the cycle, work and heat of each process, work input, work output, net work output, heat added, heat removed, MEP and cycle efficiency.

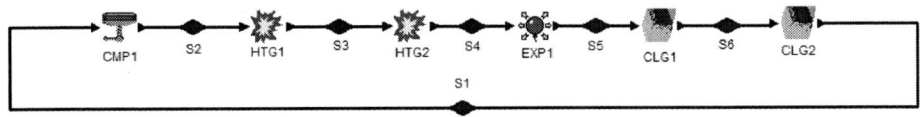

Figure E9.10.1a. Six-process internal combustion engine design.

To evaluate this design by CyclePad, we take the following steps:

1. Build
   (A) Take a compression device, two heaters, an expander and two coolers from the closed system inventory shop and connect the six devices to form the cycle as shown in Figure E9.10.1a.
   (B) Switch to analysis mode.
2. Analysis
   (A) Assume a process for each of the six processes: (a) compression device as isentropic, (b) one heater as isochoric and the other as isobaric, (c) expander as isentropic, and (d) one cooler as isochoric and the other as isobaric.
   (B) Input the given information: working fluid is air, m=0.01 kg, p_1=100 kPa, T_1=20°C, V_1=10V_2, q_{23}=600 kJ/kg, q_{34}=400 kJ/kg, and T_5=400°C.
3. Display results
   (A) Display the cycle properties results. The cycle is a heat engine. The results are: p_1=100 kPa, T_1=20°C, p_2=2512 kPa, T_2=463.2°C, p_3=5368 kPa, T_3=1300°C, p_4=5368 kPa, T_4=1699°C, p_5=124.7 kPa, T_5=400°C, p_6=100 kPa, T_6=266.6°C, Q_{12}=0, W_{12}=-3.18 kJ, Q_{23}=6 kJ, W_{23}=0, Q_{34}=4 kJ, W_{34}=1.14 kJ, Q_{45}=0, W_{45}=9.31 kJ, Q_{56}=-0.9558 kJ, W_{56}=0, Q_{61}=-2.47 kJ, W_{61}=-0.7071 kJ, W_{add}=-3.88 kJ, W_{out}=10.45 kJ, W_{net}=6.57 kJ, Q_{in}=10 kJ, Q_{out}=-3.43 kJ, MEP=448.9 kPa, and η=65.69% as shown in Figure E9.10.1b.

Design Examples   77

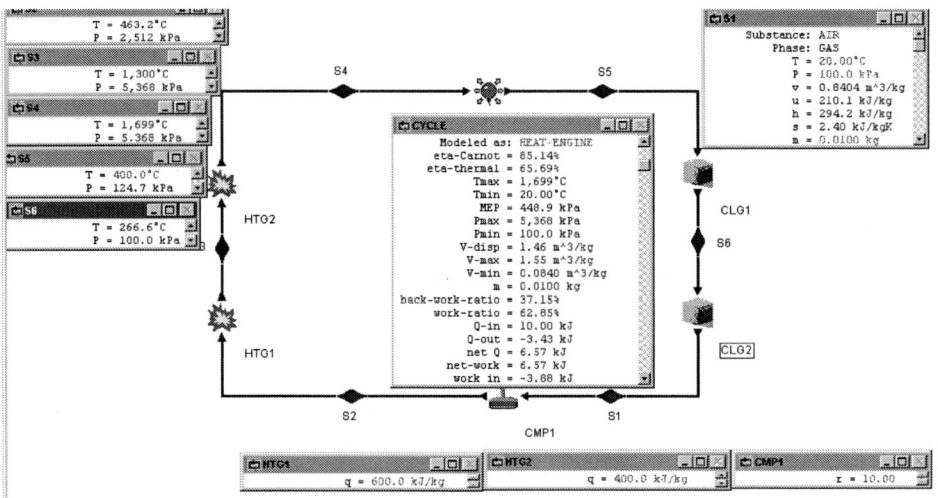

Figure E9.10.1b. Six-process internal combustion engine design results.

The T-s diagram of the cycle is shown in Figure E9.10.1c.

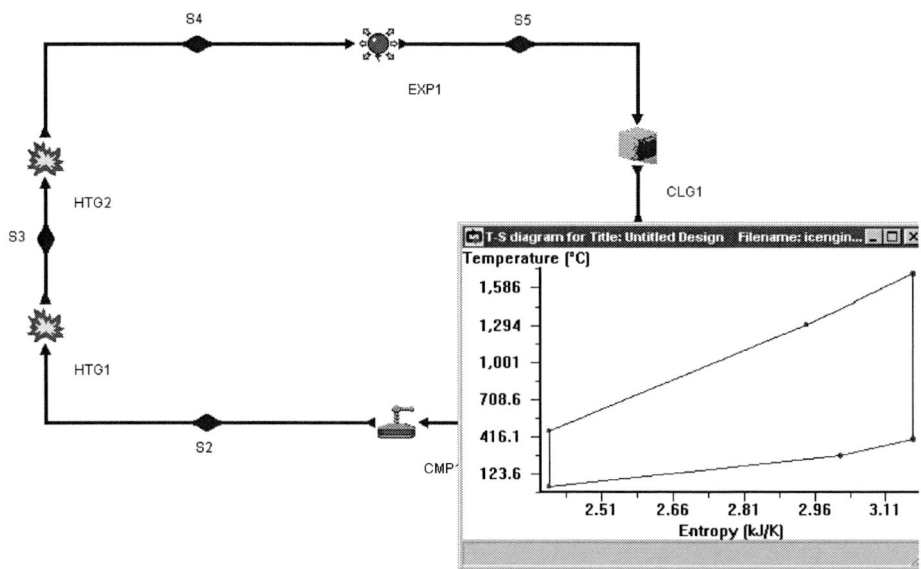

Figure E9.10.1c. T-s diagram.

The sensitivity diagram of η (cycle efficiency) versus r (compression ratio) is plotted in Figure E9.10.1d. The figure shows that the larger the compression ratio, the better the cycle efficiency. To improve the proposed engine, a larger compression ratio could be used.

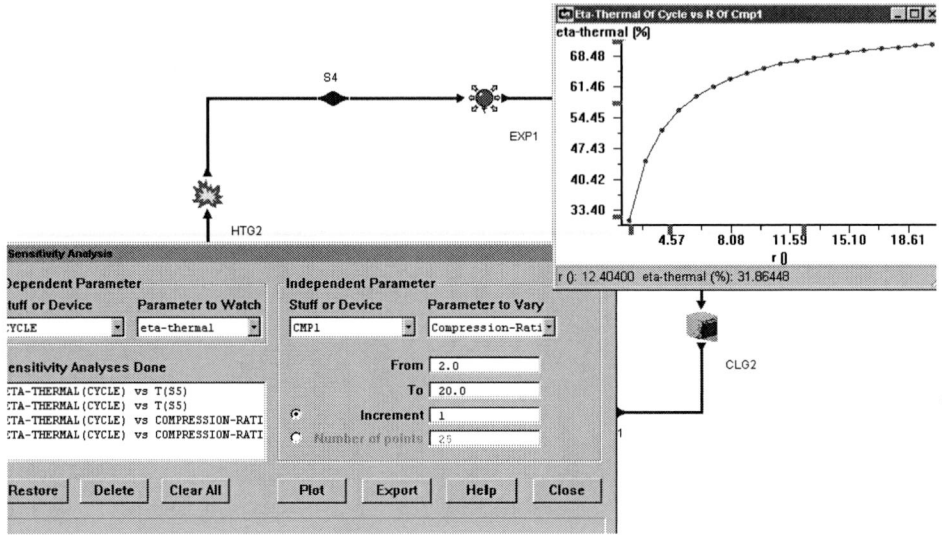

Figure E9.10.1d. Sensitivity diagram.

EXAMPLE 9.10.2.

A six-process internal combustion engine as shown in Figure E9.10.2a is proposed by a junior engineer. Air mass contained in the cylinder is 0.01 kg. The six processes are:

Process 1-2 isentropic compression
Process 2-3 isochoric heating
Process 3-4 isobaric heating
Process 4-5 isentropic expansion
Process 5-6 isobaric cooling
Process 6-1 isochoric cooling

The following information is given:

p_1=100 kPa, T_1=20°C, V_1=10V_2, q_{23}=600 kJ/kg, q_{34}=400 kJ/kg, and T_5=400°C.

Determine the pressure and temperature of each state of the cycle, work and heat of each process, work input, work output, net work output, heat added, heat removed, MEP and cycle efficiency.

Notice that process 5-6 and process 6-1 of the cycle are different from process 5-6 and process 6-1 of the cycle proposed in Example 9.10.1.

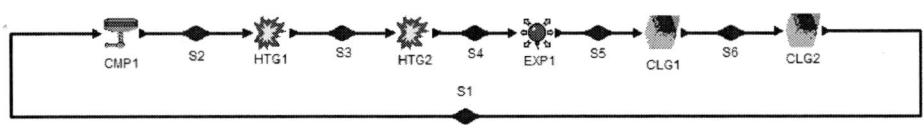

Figure E9.10.2a. Six process internal combustion engine.

To evaluate this design by CyclePad, we take the following steps:

1. Build
   (A) Take a compression device, two heaters, an expander and two coolers from the closed system inventory shop and connect the four devices to form the cycle as shown in Figure E9.10.2a.
   (B) Switch to analysis mode.
2. Analysis
   (A) Assume a process for each of the six processes: (a) compression device as isentropic, (b) one heater as isochoric and the other as isobaric, (c) expander as isentropic, and (d) one cooler as isochoric and the other as isobaric.
   (B) Input the given information: working fluid is air, m=0.01 kg, p_1=100 kPa, T_1=20°C, V_1=10V_2, q_{23}=600 kJ/kg, q_{34}=400 kJ/kg, and T_5=400°C.
3. Display results
   (A) Display the cycle properties results. The cycle is a heat engine. The results are: p_1=100 kPa, T_1=20°C, p_2=2512 kPa, T_2=463.2°C, p_3=5368 kPa, T_3=1300°C, p_4=5368 kPa, T_4=1699°C, p_5=124.7 kPa, T_5=400°C, p_6=124.7 kPa, T_6=92.42°C, Q_{12}=0, W_{12}=-3.18 kJ, Q_{23}=6 kJ, W_{23}=0, Q_{34}=4 kJ, W_{34}=1.14 kJ, Q_{45}=0, W_{45}=9.31 kJ, Q_{56}=-3.09 kJ, W_{56}=-0.8818, Q_{61}=-0.5191 kJ, W_{61}=0 kJ, W_{add}=-4.06 kJ, W_{out}=10.45 kJ, W_{net}=6.39 kJ, Q_{in}=10 kJ, Q_{out}=-3.61 kJ, MEP=436.9 kPa, and η=63.95% (see Figure E9.10.2b).

It is observed that both the cycle efficiency and MEP of the proposed cycle are less than those of the proposed cycle given by Example 9.10.1.

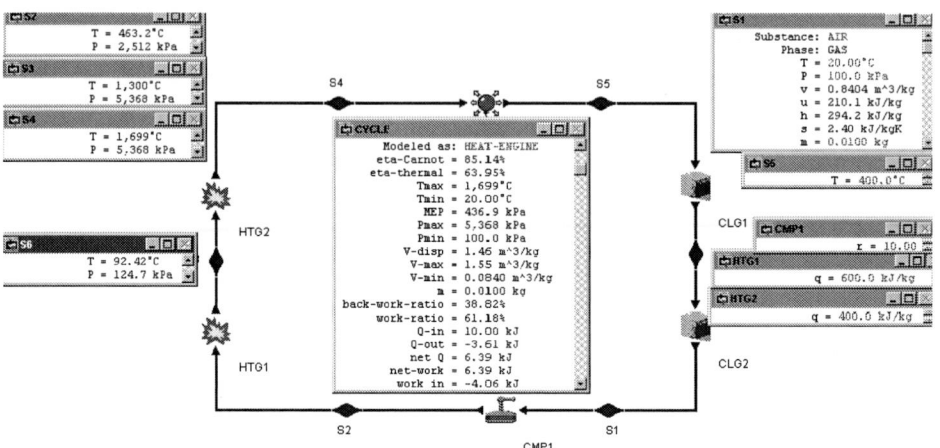

Figure E9.10.2b. Six-process internal combustion engine design results.

The T-s diagram and sensitivity diagram of η (cycle efficiency) versus r (compression ratio) is plotted in Figure E9.10.2c.

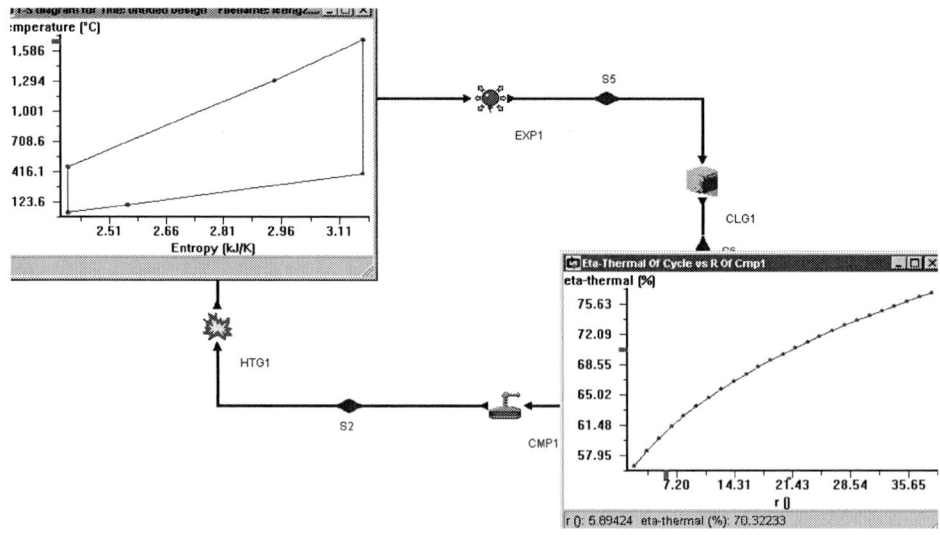

Figure E9.10.2c. T-s diagram and sensitivity diagram of η (cycle efficiency) vs r (compression ratio).

Design Examples

EXAMPLE 9.10.3.

Adding a turbo-charger and a pre-cooler to a Dual cycle is proposed as shown in Figure E9.10.3a. The cylinder volume of the engine is 0.01 m³. Evaluate the proposed cycle. The basic Dual cycle and the proposed turbo-charger and pre-cooler Dual cycle information is:

Basic Dual cycle:
$p_1=p_2=p_3=p_8=101$ kPa, $T_1=T_2=T_3=T_8=15°C$, $V_3=0.01$ m³, r(compression ratio)=10, and $q_{45}=q_{56}=300$ kJ/kg.

Turbo-charger and pre-cooler Dual cycle

$p_1=p_8=101$ kPa, $T_1=T_8=15°C$, $p_2=150$ kPa, $T_3=15°C$, $V_3=0.01$ m³, r(compression ratio)=10, and $q_{45}=q_{56}=300$ kJ/kg.

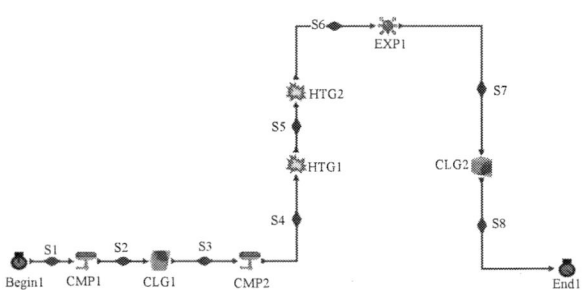

Figure E9.10.3a. Turbo-charger and pre-cooler Dual cycle.

To evaluate this proposed cycle by CyclePad, we take the following steps:

1. Build
   (A) Take two compression devices, two heaters, an expander and two coolers from the closed system inventory shop and connect the devices to form the cycle.
   (B) Switch to analysis mode.

2. Analysis
    (A) Assume a process for each of the seven processes: (a) compression devices as isentropic, (b) one heater as isochoric and the other as isobaric, (c) expander as isentropic, and (d) one cooler as isochoric and the other as isobaric.
    (B) Input the given information (see Figure E9.10.3b): working fluid is air, $p_1=p_8=101$ kPa, $T_1=T_8=15°C$, $p_2=150$ kPa, $T_3=15°C$, $V_3=0.01$ m^3, r(compression ratio)=10, and $q_{45}=q_{56}=300$ kJ/kg.
3. Display results
    (A) Display the cycle properties results. The cycle is a heat engine. The results are: $p_1=100$ kPa, $T_1=15°C$, $p_2=150$ kPa, $T_2=49.47°C$, $p_3=150$ kPa, $T_3=15°C$, $p_4=3768$ kPa, $T_4=450.7°C$, $p_5=5947$ kPa, $T_5=869.2°C$, $p_6=5947$ kPa, $T_6=1168°C$, $p_7=188.4$ kPa, $T_7=264.4°C$, $Q_{12}=0$, $W_{12}=-0.4486$ kJ, $Q_{23}=-0.6281$ kJ, $W_{23}=-0.1795$, $W_{34}=-5.67$ kJ, $Q_{45}=0$, $W_{45}=11.76$ kJ, $Q_{56}=-3.25$ kJ, $W_{add}=-6.30$ kJ, $W_{out}=13.32$ kJ, $W_{net}=7.02$ kJ, $Q_{in}=10.89$ kJ, $Q_{out}=-3.87$ kJ, MEP=506.8 kPa, and $\eta=64.44\%$ as shown in Figure E9.10.3c.

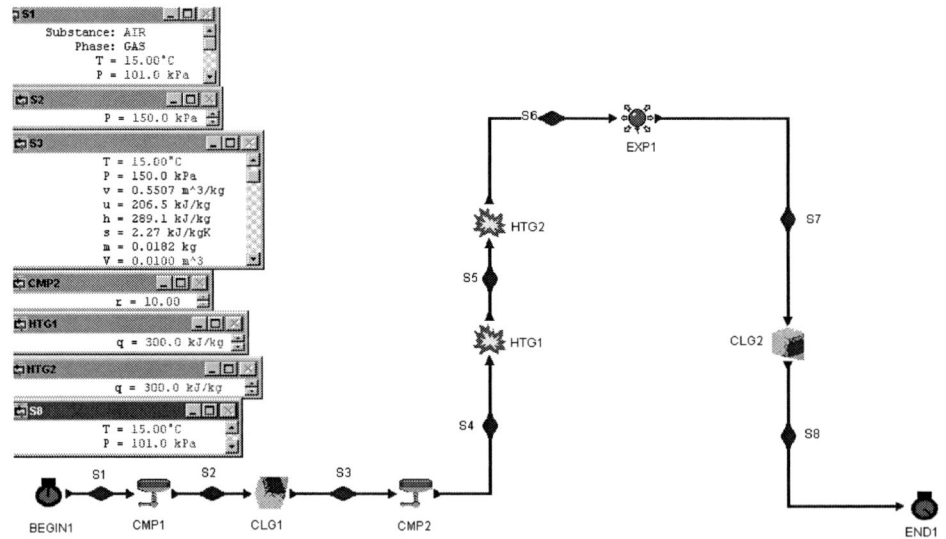

Figure E9.10.3b. Turbo-charger and pre-cooler Dual cycle input.

Design Examples 83

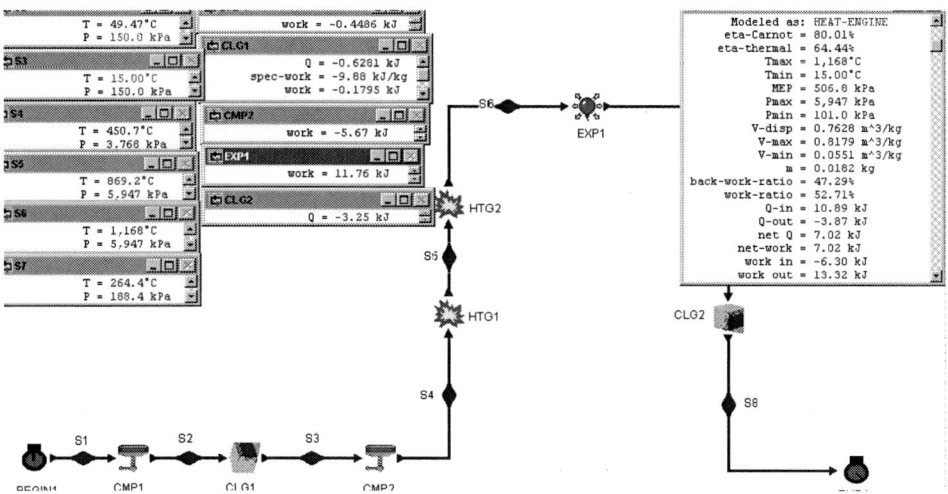

Figure E9.10.3c. Turbo-charger and pre-cooler Dual cycle result.

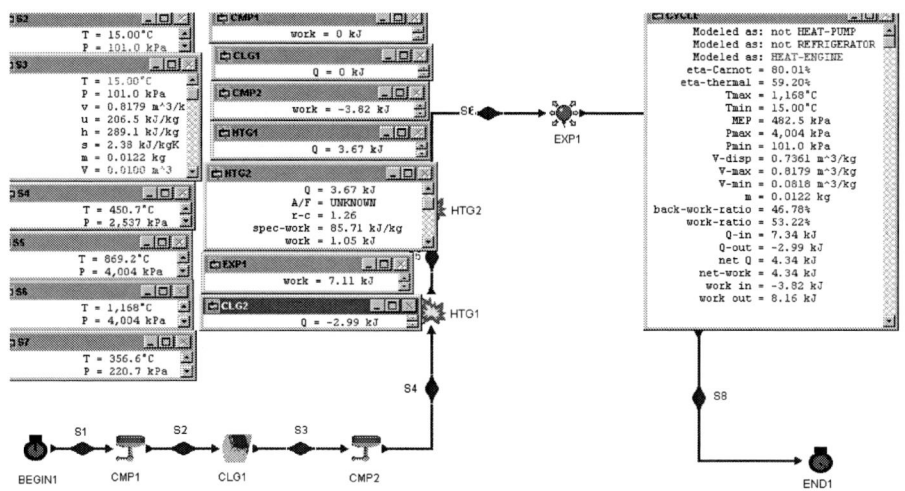

Figure E9.10.3d. Dual cycle without result turbo-charger and pre-cooler.

For the Dual cycle without turbo-charger and pre-cooler, the input are: $p_1=p_2=p_3=p_8=101$ kPa, $T_1=T_2=T_3=T_8=15°C$, $V_3=0.01$ m^3, r(compression ratio)=10, and $q_{45}=q_{56}=300$ kJ/kg as shown in Figure E9.10.3d.

The output results are: $p_1=p_2=p_3=101$ kPa, $T_1=T_2=T_3=15°C$, $p_4=2537$ kPa, $T_4=450.7°C$, $p_5=4004$ kPa, $T_5=869.2°C$, $p_6=4004$ kPa, $T_6=1168°C$, $p_7=220.7$ kPa, $T_7=355.6°C$, $Q_{12}=0$, $W_{12}=-0$ kJ, $Q_{23}=-0$ kJ, $W_{23}=-0$ kJ, $W_{34}=-3.82$ kJ, $Q_{45}=0$,

W_{45}=7.11 kJ, Q_{56}=-2.99 kJ, W_{add}=-3.82 kJ, W_{out}=8.16 kJ, W_{net}=4.34 kJ, Q_{in}=7.34 kJ, Q_{out}=-2.94 kJ, MEP=482.5 kPa, and η=59.20% as shown in Figure E9.10.3d.

It is observed that both the cycle efficiency and MEP of the proposed cycle are better than those of the Dual cycle without turbo-charger and pre-cooler.

HOMEWORK 9.10. DESIGN

1. An imaginary ideal gas cycle is made by three processes. Air at 1 bar, 0.5 m^3 and 300 K is compressed in a constant volume compression process (Process 1-2, zero work) to 6 bar; air is expanded in an isentropic expansion process Process (2-3) to 1 bar; and air is compressed in an isobaric compression process (Process 3-1) to 0.5 m^3. Find the heat added, heat removed, work added, work produced, net work produced, and cycle efficiency.
ANSWER: 625 kJ, -454.3 kJ, -129.8 kJ, 300.5 kJ, 170.7 kJ, 27.31%.

2. An imaginary ideal gas cycle is made by three processes. Air at 1 bar, 0.5 m^3 and 300 K is compressed in a constant volume compression process (Process 1-2, zero work) to 8 bar; air is expanded in an isentropic expansion process Process (2-3) to 1 bar; and air is compressed in an isobaric compression process (Process 3-1) to 0.5 m^3. Find the heat added, heat removed, work added, work produced, net work produced, and cycle efficiency.
ANSWER: 875 kJ, -597.7 kJ, -170.8 kJ, 448.0 kJ, 277.1 kJ, 31.67%.

3. An imaginary ideal gas cycle is made by three processes. Air at 14.7 psia, 12 ft^3 and 80°F is compressed in a constant volume compression process (Process 1-2, zero work) to 120 psia; air is expanded in an isentropic expansion process Process (2-3) to 14.7 psia; and air is compressed in an isobaric compression process (Process 3-1) to 12 ft^3. Find the heat added, heat removed, work added, work produced, net work produced, and cycle efficiency.
ANSWER: 584.6 Btu, -397.7 Btu, -113.6 Btu, 300.6 Btu, 186.9 Btu, 31.98%.

4. An imaginary ideal gas cycle is made by three processes. Air at 14.7 psia, 12 ft^3 and 80°F is compressed in a constant volume compression process (Process 1-2, zero work) to 100 psia; air is expanded in an isentropic expansion process Process (2-3) to 14.7 psia; and air is compressed in an isobaric compression process (Process 3-1) to 12 ft^3. Find the heat added,

Design Examples 85

heat removed, work added, work produced, net work produced, and cycle efficiency.
ANSWER: 473.6 Btu, -335.2 Btu, -95.77 Btu, 234.2 Btu, 138.4 Btu, 29.23%.

5. An imaginary ideal gas engine cycle is made by three processes. Air at 1 bar, 0.05 m^3 and 300 K is compressed in a constant volume compression process (Process 1-2, zero work) to 6 bar; air is expanded in an isobaric expansion process Process (2-3) to 0.9 m^3; and 100 kJ of heat is removed from air in a cooling process (Process 3-1) to 1 bar, 0.05 m^3 and 300 K. Find the heat added, heat removed, work added, work produced, net work produced, and cycle efficiency.
ANSWER: 146.5 kJ, -100.0 kJ, 0 kJ, 46.50 kJ, 46.50 kJ, 31.74%.

6. An imaginary ideal gas engine cycle is made by three processes. Air at 1 bar, 0.05 m^3 and 300 K is compressed in a constant volume compression process (Process 1-2, zero work) to 8 bar; air is expanded in an isobaric expansion process Process (2-3) to 0.9 m^3; and 100 kJ of heat is removed from air in a cooling process (Process 3-1) to 1 bar, 0.05 m^3 and 300 K. Find the heat added, heat removed, work added, work produced, net work produced, and cycle efficiency.
ANSWER: 199.5 kJ, -100.0 kJ, 0 kJ, 99.50 kJ, 99.50 kJ, 49.87%.

7. An imaginary ideal gas engine cycle is made by three processes. Air at 1 bar, 0.05 m^3 and 300 K is compressed in a constant volume compression process (Process 1-2, zero work) to 8 bar; air is expanded in an isobaric expansion process Process (2-3) to 0.9 m^3; and 120 kJ of heat is removed from air in a cooling process (Process 3-1) to 1 bar, 0.05 m^3 and 300 K. Find the heat added, heat removed, work added, work produced, net work produced, and cycle efficiency.
ANSWER: 199.3 kJ, -120.0 kJ, 0 kJ, 79.34 kJ, 79.34 kJ, 39.80%.

8. An imaginary ideal gas engine cycle is made by three processes. Air at 14.7 psia, 2.5 ft^3 and 80°F is compressed in a constant volume compression process (Process 1-2, zero work) to 120 psia; air is expanded in an isobaric expansion process Process (2-3) to 0.9 m^3; and 120 Btu of heat is removed from air in a cooling process (Process 3-1) to 14.7 psia, 2.5 ft^3 and 80°F. Find the heat added, heat removed, work added, work produced, net work produced, and cycle efficiency.
ANSWER: 165.3 Btu, -120.0 Btu, 0 Btu, 45.32 Btu, 45.32 Btu, 27.41%.

9. An imaginary ideal gas engine cycle is made by three processes. Heat in the amount of 30 kJ is added to air at 1 bar, 0.005 m^3 and 300 K in a heating process (Process 1-2) to 6 bar and 0.015 m^3; air is expanded in a

constant volume process Process (2-3, zero work) to 1 bar; and air is compressed in an isobaric process (Process 3-1) to 1 bar, 0.005 m³ and 300 K. Find the heat added, heat removed, work added, work produced, net work produced, and cycle efficiency.
ANSWER: 30.0 kJ, -22.25 kJ, -1 kJ, 8.75 kJ, 7.75 kJ, 25.83%.

10. An imaginary ideal gas engine cycle is made by three processes. Heat in the amount of 30 kJ is added to air at 1 bar, 0.005 m³ and 300 K in a heating process (Process 1-2) to 8 bar and 0.015 m³; air is expanded in a constant volume process Process (2-3, zero work) to 1 bar; and air is compressed in an isobaric process (Process 3-1) to 1 bar, 0.005 m³ and 300 K. Find the heat added, heat removed, work added, work produced, net work produced, and cycle efficiency.
ANSWER: 30.0 kJ, -29.75 kJ, -1 kJ, 1.25 kJ, 0.25 kJ, 0.8333%.

11. An imaginary ideal gas engine cycle is made by three processes. Heat in the amount of 50 kJ is added to air at 100 kPa, 0.005 m³ and 25°C in a heating process (Process 1-2) to 600 kPa and 0.015 m³; air is expanded in a constant volume process Process (2-3, zero work) to 1 bar; and air is compressed in an isobaric process (Process 3-1) to 100 kPa, 0.005 m³ and 25°C. Find the heat added, heat removed, work added, work produced, net work produced, and cycle efficiency.
ANSWER: 50.0 kJ, -29.75 kJ, -1 kJ, 21.25 kJ, 20.25 kJ, 40.50%.

12. An imaginary ideal gas engine cycle is made by three processes. Heat in the amount of 40 Btu is added to air at 14.7 psia, 0.4 ft³ and 77°F in a heating process (Process 1-2) to 100 psia and 0.4 ft³; air is expanded in a constant volume process Process (2-3, zero work) to 14.7 psia; and air is compressed in an isobaric process (Process 3-1) to 14.7 psia, 0.4 ft³ and 77°F. Find the heat added, heat removed, work added, work produced, net work produced, and cycle efficiency.
ANSWER: 40.0 Btu, -18.64 Btu, -0.8162 Btu, 22.17 Btu, 21.36 kJ, 53.39%.

13. Air at 1 bar, 0.1 m³ and 300 K is (A) heated reversibly at constant volume until its pressure is two times of its initial value; (B) heated reversibly at constant pressure until the volume is two times of its initial value; (C) cooled reversibly at constant volume until its pressure returns to its initial value; and (D) cooled reversibly at constant pressure to the initial state. Find the heat added, heat removed, work added, work produced, net work produced, and cycle efficiency.
ANSWER: 95.0 kJ, -85.0 kJ, -10.0 kJ, 20.0 kJ, 10.0 kJ, 10.53%.

Design Examples 87

14. Air at 1 bar, 0.1 m³ and 300 K is (A) heated reversibly at constant volume until its pressure is two times of its initial value; (B) heated reversibly at constant pressure until the volume is three times of its initial value; (C) cooled reversibly at constant volume until its pressure returns to its initial value; and (D) cooled reversibly at constant pressure to the initial state. Find the heat added, heat removed, work added, work produced, net work produced, and cycle efficiency.
ANSWER: 165.0 kJ, -145.0 kJ, -20.0 kJ, 40.0 kJ, 20.0 kJ, 12.12%.

15. Air at 100 kPa, 0.05 m³ and 30°C is (A) heated reversibly at constant volume until its pressure is two times of its initial value; (B) heated reversibly at constant pressure until the volume is two times of its initial value; (C) cooled reversibly at constant volume until its pressure returns to its initial value; and (D) cooled reversibly at constant pressure to the initial state. Find the heat added, heat removed, work added, work produced, net work produced, and cycle efficiency.
ANSWER: 187.5 kJ, -162.5 kJ, -25.0 kJ, 50.0 kJ, 25.0 kJ, 13.33%.

16. Helium at 100 kPa, 0.05 m³ and 30°C is (A) heated reversibly at constant volume until its pressure is two times of its initial value; (B) heated reversibly at constant pressure until the volume is two times of its initial value; (C) cooled reversibly at constant volume until its pressure returns to its initial value; and (D) cooled reversibly at constant pressure to the initial state. Find the heat added, heat removed, work added, work produced, net work produced, and cycle efficiency.
ANSWER: 132.1 kJ, -107.1 kJ, -25.0 kJ, 50.0 kJ, 25.0 kJ, 18.93%.

17. Helium at 14.7 psia, 1 ft³ and 80°F is (A) heated reversibly at constant volume until its pressure is 40 psia; (B) heated reversibly at constant pressure until the volume is 5 ft³; (C) cooled reversibly at constant volume until its pressure returns to its initial value; and (D) cooled reversibly at constant pressure to the initial state. Find the heat added, heat removed, work added, work produced, net work produced, and cycle efficiency.
ANSWER: 80.8 Btu, -62.07 Btu, -10.88 Btu, 29.61 Btu, 18.73 kJ, 23.18%.

18. Air at 100 kPa, 0.1 m³ and 300 K undergoes the following cycle of operations: (A) heating at constant volume from 300 K to 600 K; (B) isothermal expansion to 0.3 m³; (C) cooling at constant volume from 600 K to 300 K; (D) isothermal compression to the initial state. Find the heat added, heat removed, work added, work produced, net work produced, and cycle efficiency.

ANSWER: 46.97 kJ, -35.99 kJ, -10.99 kJ, 21.97 kJ, 10.99 kJ, 23.39%.

19. Air at 100 kPa, 0.1 m^3 and 300 K undergoes the following cycle of operations: (A) heating at constant volume from 300 K to 800 K; (B) isothermal expansion to 0.3 m^3; (C) cooling at constant volume from 800 K to 300 K; (D) isothermal compression to the initial state. Find the heat added, heat removed, work added, work produced, net work produced, and cycle efficiency.
ANSWER: 70.96 kJ, -52.65 kJ, -10.99 kJ, 29.30 kJ, 18.31 kJ, 25.80%.

20. Carbon dioxide at 100 kPa, 0.1 m^3 and 300 K undergoes the following cycle of operations: (A) heating at constant volume from 300 K to 800 K; (B) isothermal expansion to 0.3 m^3; (C) cooling at constant volume from 800 K to 300 K; (D) isothermal compression to the initial state. Find the heat added, heat removed, work added, work produced, net work produced, and cycle efficiency.
ANSWER: 86.77 kJ, -68.46 kJ, -10.99 kJ, 29.30 kJ, 18.31 kJ, 21.10%.

21. Helium at 100 kPa, 0.1 m^3 and 300 K undergoes the following cycle of operations: (A) heating at constant volume from 300 K to 800 K; (B) isothermal expansion to 0.3 m^3; (C) cooling at constant volume from 800 K to 300 K; (D) isothermal compression to the initial state. Find the heat added, heat removed, work added, work produced, net work produced, and cycle efficiency.
ANSWER: 54.17 kJ, -35.86 kJ, -10.99 kJ, 29.30 kJ, 18.31 kJ, 33.80%.

22. Air at 100 kPa, 1 m^3 and 300 K is heated to 1200 K at constant pressure. It then expands isentropically until the temperature falls to 900 K, is then cooled at constant volume to 300 K, and compressed isothermally to its initial state.

23. Find the heat added, heat removed, work added, work produced, net work produced, and cycle efficiency.
ANSWER: 1050 kJ, -710.5 kJ, -210.5 kJ, 550 kJ, 339.5 kJ, 32.33%.

24. Air at 100 kPa, 1 m^3 and 300 K is heated to 1600 K at constant pressure. It then expands isentropically until the temperature falls to 1000 K, is then cooled at constant volume to 300 K, and compressed isothermally to its initial state.

25. Find the heat added, heat removed, work added, work produced, net work produced, and cycle efficiency.
ANSWER: 1517 kJ, -868.2 kJ, -284.9 kJ, 933.3 kJ, 648.4 kJ, 42.75%.

26. Helium at 100 kPa, 1 m^3 and 300 K is heated to 1600 K at constant pressure. It then expands isentropically until the temperature falls to 1000

K, is then cooled at constant volume to 300 K, and compressed isothermally to its initial state.

27. Find the heat added, heat removed, work added, work produced, net work produced, and cycle efficiency.
ANSWER: 1080 kJ, -585.8 kJ, -237.5 kJ, 731.8 kJ, 494.3 kJ, 45.76%.

28. Carbon dioxide at 100 kPa, 1 m^3 and 300 K is heated to 1600 K at constant pressure. It then expands isentropically until the temperature falls to 1000 K, is then cooled at constant volume to 300 K, and compressed isothermally to its initial state. Find the heat added, heat removed, work added, work produced, net work produced, and cycle efficiency.
ANSWER: 1928 kJ, -1134 kJ, -329.5 kJ, 1123 kJ, 793.5 kJ, 41.17%.

29. Air at 200 kPa, 0.5 m^3 and 300 K is heated to 1800 K at constant pressure. It then expands isentropically until the temperature falls to 300 K, is then compressed isothermally to its initial state. Find the heat added, heat removed, work added, work produced, net work produced, and cycle efficiency.
ANSWER: 1750 kJ, -627.1 kJ, -627.1 kJ, 1750 kJ, 1123 kJ, 64.16%.

30. Air at 200 kPa, 0.5 m^3 and 300 K is heated to 1500 K at constant pressure. It then expands isentropically until the temperature falls to 300 K, is then compressed isothermally to its initial state. Find the heat added, heat removed, work added, work produced, net work produced, and cycle efficiency.
ANSWER: 1400 kJ, -563.3 kJ, -563.3 kJ, 1400 kJ, 836.7 kJ, 59.76%.

31. Helium at 200 kPa, 0.5 m^3 and 300 K is heated to 1500 K at constant pressure. It then expands isentropically until the temperature falls to 300 K, is then compressed isothermally to its initial state. Find the heat added, heat removed, work added, work produced, net work produced, and cycle efficiency.
ANSWER: 997.0 kJ, -401.2 kJ, -401.2 kJ, 997.0 kJ, 595.9 kJ, 59.76%.

32. Helium at 20 psia, 1 ft^3 and 80°F is heated to 2000°F at constant pressure. It then expands isentropically until the temperature falls to 80°F, is then compressed isothermally to its initial state. Find the heat added, heat removed, work added, work produced, net work produced, and cycle efficiency.
ANSWER: 32.82 Btu, -13.99 Btu, -13.99 Btu, 32.82 Btu, 18.83 kJ, 57.37%.

33. Carbon dioxide at 20 psia, 1 ft^3 and 80°F is heated to 2000°F at constant pressure. It then expands isentropically until the temperature falls to 80°F,

is then compressed isothermally to its initial state. Find the heat added, heat removed, work added, work produced, net work produced, and cycle efficiency.
ANSWER: 58.58 Btu, -24.97 Btu, -24.97 Btu, 58.58 Btu, 33.60 kJ, 57.37%.

34. A six-process internal combustion engine as shown in Figure E9.10.2a is proposed by a junior engineer. Air mass contained in the cylinder is 0.01 kg. The six processes are:
    Process 1-2   isentropic compression
    Process 2-3   isochoric heating
    Process 3-4   isobaric heating
    Process 4-5   isentropic expansion
    Process 5-6   isobaric cooling
    Process 6-1   isochoric cooling
    The following information is given as shown in Figure E9.10.2b:
    p_1=14.7 psia, T_1=60°F, V_1=8V_2, q_{23}=130 Btu/lbm, q_{34}=170 Btu/lbm, and T_5=750°F.
    Determine the work input, work output, net work output, heat added, heat removed, MEP and cycle efficiency.
    ANSWER: -2.61 Btu, 5.95 Btu, 3.33 Btu, 6 Btu, -2.67 Btu, 62.61 psia, 55.56%.

35. A six-process internal combustion engine as shown in Figure E9.10.2a is proposed by a junior engineer. Air mass contained in the cylinder is 0.02 lbm. The six processes are:
    Process 1-2   isentropic compression
    Process 2-3   isochoric heating
    Process 3-4   isobaric heating
    Process 4-5   isentropic expansion
    Process 5-6   isobaric cooling
    Process 6-1   isochoric cooling
    The following information is given as shown in Figure E9.10.2b:
    p_1=100 kPa, T_1=15°C, V_1=8V_2, q_{23}=300 kJ/kg, q_{34}=400 kJ/kg, and T_5=400°C.
    Determine the pressure and temperature of each state of the cycle, work and heat of each process, work input, work output, net work output, heat added, heat removed, MEP and cycle efficiency.
    ANSWER: [(100 kPa, 15°C), (1838 kPa, 388.8°C), (3000 kPa, 807.4°C), (3000 kPa, 1206°C), (190.7 kPa, 400°C), (190.7 kPa, 276.4°C)], [(-2.68 kJ, 0 kJ), (0 kJ, 3 kJ), (1.14 kJ, 4 kJ), (5.78 kJ, 0 kJ), (-0.3543 kJ, -1.24

kJ), (0 kJ, -1.87 kJ)], -3.03 kJ, 6.92 kJ, 3.89 kJ, 7 kJ, -3.11 kJ, 427.7 kPa, 55.52%.

36. Adding a turbo-charger and a pre-cooler to a Dual cycle is proposed as shown in Figure E9.10.3a. The cylinder volume of the engine is 0.01 m^3. Evaluate the proposed cycle. The basic Dual cycle and the proposed turbo-charger and pre-cooler Dual cycle information are:
Basic Dual cycle:
$p_1=p_2=p_3=p_8=101$ kPa, $T_1=T_2=T_3=T_8=15°C$, $V_3=0.01$ m^3, r(compression ratio)=8, and $q_{45}=q_{56}=300$ kJ/kg.
Turbo-charger and pre-cooler Dual cycle
$p_1=p_8=101$ kPa, $T_1=T_8=15°C$, $p_2=120$ kPa, $T_3=15°C$, $V_3=0.01$ m^3, r(compression ratio)=10, and $q_{45}=q_{56}=300$ kJ/kg.
ANSWER: [Dual cycle--η=59.20%, MEP=482.5 kPa, m=0.0122 kg, Qi=7.34 kJ, Qo=-2.99 kJ, Wi=-3.82 kJ, Wo=8.16 kJ, Wn=4.34 kJ], [Proposed cycle--η=61.78%, MEP=494.8 kPa, m=0.0145 kg, Qi=8.72 kJ, Qo=-3.33 kJ, Wi=-4.75 kJ, Wo=10.13 kJ, Wn=5.38 kJ].

37. Adding a turbo-charger and a pre-cooler to a Dual cycle is proposed as shown in Figure E9.10.3a. The cylinder volume of the engine is 0.3 ft^3. Evaluate the proposed cycle. The basic Dual cycle and the proposed turbo-charger and pre-cooler Dual cycle information are:
Basic Dual cycle:
$p_1=p_2=p_3=p_8=14.7$ psia, $T_1=T_2=T_3=T_8=60°F$, $V_3=0.3$ ft^3, r(compression ratio)=10, and $q_{45}=q_{56}=130$ Btu/lbm.
Turbo-charger and pre-cooler Dual cycle
$p_1=p_8=14.7$ psia, $T_1=T_8=60°F$, $p_2=20$ psia, $T_3=60°F$, $V_3=0.3$ ft^3, r(compression ratio)=10, and $q_{45}=q_{56}=130$ Btu/lbm.

38. ANSWER: [Dual cycle--η=59.19%, MEP=70.64 psia, m=0.0229 lbm, Qi=5.96 Btu, Qo=-2.43 Btu, Wi=-3.08 Btu, Wo=6.61 Btu, Wn=3.53 Btu], [Proposed cycle--η=63.48%, MEP=73.59 psia, m=0.0312 lbm, Qi=8.11 Btu, Qo=-2.96 Btu, Wi=-4.55 Btu, Wo=9.70 Btu, Wn=5.15 Btu]].

39. Adding a turbo-charger and a pre-cooler to a Dual cycle is proposed as shown in Figure E9.10.3a. The cylinder volume of the engine is 0.3 ft^3. Evaluate the proposed cycle. The basic Dual cycle and the proposed turbo-charger and pre-cooler Dual cycle information are:
Basic Dual cycle:
$p_1=p_2=p_3=p_8=14.7$ psia, $T_1=T_2=T_3=T_8=60°F$, $V_3=0.3$ ft^3, r(compression ratio)=8, and $q_{45}=q_{56}=130$ Btu/lbm.
Turbo-charger and pre-cooler Dual cycle

$p_1=p_8=14.7$ psia, $T_1=T_8=60°F$, $p_2=20$ psia, $T_3=60°F$, $V_3=0.3$ ft^3, r(compression ratio)=8, and $q_{45}=q_{56}=130$ Btu/lbm.

ANSWER: [Dual cycle--η=55.32%, MEP=67.91 psia, m=0.0229 lbm, Qi=5.96 Btu, Qo=-2.66 Btu, Wi=-2.65 Btu, Wo=5.95 Btu, Wn=3.30 Btu], [Proposed cycle--η=60.06%, MEP=71.04 psia, m=0.0312 lbm, Qi=8.11 Btu, Qo=-3.24 Btu, Wi=-3.96 Btu, Wo=8.83 Btu, Wn=4.87 Btu]].

Chapter 11

SUMMARY

Heat engines that use gases as the working fluid in a closed system model were discussed in this chapter. Otto cycle, Diesel, Miller, and Dual cycle are internal combustion engines. Stirling cycle is an external combustion engine.

The Otto cycle is a spark-ignition reciprocating engine made of an isentropic compression process, a constant volume combustion process, an isentropic expansion process, and a constant volume cooling process. The thermal efficiency of the Otto cycle depends on its compression ratio. The compression ratio is defined as $r=V_{max}/V_{min}$. The Otto cycle efficiency is limited by the compression ratio because of the engine knock problem.

The Diesel cycle is a compression-ignition reciprocating engine made of an isentropic compression process, a constant pressure combustion process, an isentropic expansion process, and a constant volume cooling process. The thermal efficiency of the Otto cycle depends on its compression ratio and cut-off ratio. The compression ratio is defined as $r=V_{max}/V_{min}$. The cut-off ratio is defined as $r_{cutoff}=V_{combustion\ off}/V_{min}$.

The Dual cycle involves two heat addition processes, one at constant volume and one at constant pressure. It behaves more like an actual cycle than either Otto or Diesel cycle.

The Lenoir cycle was the first commercially successful internal combustion engine.

The Stirling cycle and Wicks cycle are attempt to achieve the Carnot efficiency.

The Miller cycle uses variable valve timing for compression ratio control to improve the performance of internal combustion engines.

REFERENCES

Balmer, Robert T., *Thermodynamics*, West Publishing Co., New York, 1990.
Balzhiser, Richard E. and Michael R. Samuels, *Engineering Thermodynamics*, Prentice-Hall Publishing Co., New Jersey, 1977.
Black, William Z, and James G. Hartley, *Thermodynamics*, Third Edition, Harper-Collins College Publishers, New York, 1998.
Cengel, Yunus A. and Michael A. Boles, *Thermodynamics: An Engineering Approach,* Third Edition, McGraw-Hill Publishing Co., New York, 1998.
Granet, Irving and Maurice Bluestein, *Thermodynamics and Heat Power*, Sixth Edition, Prentice-Hall Publishing Co., New Jersey, 2000.
Haberman, William, L. and James E.A. John, *Engineering Thermodynamics*, Allen and Bacon, Inc., Boston, 1980.
Huang, Francis F., Engineering *Thermodynamics, Fundamentals and Applications*, MacMillian Publishing Co., New York, 1976.
Moran, Michael, J. and Howard N. Shapiro, *Fundamentals of Thermodynamics*, Fourth Edition, John Wiley and Sons Publishing Co., New York, 2000.
Rogers, G.F.C. and Y.R. Mayhew, *Engineering Thermodynamics, Work and Heat Transfer,* Fourth Edition, Addison Wesley Longman Ltd, Essex, England, 1992.
Sonntag, Richard E., Claus Borgnakke, and Gordon J. Van Wylen, *Fundamentals of Thermodynamics*, Fifth Edition, John Wiley and Sons Publishing Co., New York, 1998.
Wark, Kenneth, Jr. and Donald E. Richards, *Thermodynamics*, Sixth Edition, McGraw-Hill Publishing Co., New York, 1999.
Wu, Chih, Lingen Chen and Jincan Chen, *Recent Advances in Finite Time Thermodynamics*, Nova Science Publishing Co., New York, 1999.

Wu, Chih, *Thermodynamic cycles: computer-aided design and optimization*, Marcel-Dekker Publ. Inc., New York, 2004.

INDEX

A

adiabatic, 6, 63
aid, 11
air, 1, 3, 4, 5, 6, 7, 8, 9, 10, 12, 13, 14, 15, 16, 17, 21, 22, 23, 24, 25, 26, 27, 28, 29, 30, 31, 32, 34, 35, 36, 39, 41, 42, 43, 44, 45, 46, 55, 59, 60, 61, 62, 63, 65, 70, 76, 79, 82, 84, 85, 86
angular velocity, 17
application, 5
assumptions, 31
atmosphere, 9, 10, 25, 26, 27, 31, 32, 59, 60, 62
atmospheric pressure, 8, 11, 24
automobiles, 1, 59

B

barrier, 4
behavior, 75
benefits, 61
burning, 63

C

Carbon, 55, 73, 88, 89
carbon dioxide, 55
Carnot, vii, 12, 31, 52, 54, 63, 75, 93
combustion, vii, 1, 5, 6, 8, 11, 12, 13, 14, 15, 16, 17, 19, 22, 23, 24, 29, 31, 32, 34, 35, 36, 37, 39, 40, 41, 43, 45, 49, 51, 53, 59, 60, 61, 62, 63, 75, 76, 77, 78, 79, 80, 90, 93
combustion chamber, 6, 8, 12, 22, 23, 24, 29, 34, 35, 36, 39, 40, 45, 51, 53, 59, 60, 61, 62
combustion processes, 41
components, 17, 19
consumption, 45
control, vii, 2, 18, 59, 93
cooling, vii, 10, 13, 24, 25, 26, 27, 29, 30, 35, 50, 57, 60, 61, 62, 67, 69, 72, 73, 75, 78, 85, 87, 88, 90, 93
cooling process, vii, 10, 13, 25, 26, 27, 29, 30, 35, 50, 60, 61, 62, 85, 93
CyclePad, 6, 7, 21, 23, 39, 45, 51, 53, 64, 70, 75, 76, 79, 81
cycles, 41, 67, 75

D

density, 8, 24
desires, 10
detonation, 28
diesel, 32
displacement, 4, 16
duration, 19, 57

E

energy, 52, 75
energy transfer, 52
engines, vii, 4, 11, 21, 37, 41, 57, 59, 63, 75, 93
environment, 12, 29, 63

F

flow, 10, 12, 59
flow rate, 59
fluid, vii, 3, 4, 6, 8, 13, 15, 17, 21, 22, 23, 34, 39, 44, 45, 50, 51, 53, 55, 65, 70, 75, 76, 79, 82, 93
fuel, 1, 4, 5, 11, 16, 17, 18, 19, 21, 28, 45, 59, 63
fuel flow rate, 59

G

gas, 1, 10, 12, 17, 29, 59, 61, 75, 84, 85, 86
gases, vii, 5, 43, 93
gasoline, 1, 4, 12, 15, 16

H

heat, vii, 1, 2, 3, 4, 5, 6, 7, 8, 9, 10, 11, 12, 13, 14, 15, 16, 19, 20, 21, 22, 23, 24, 25, 26, 27, 29, 30, 31, 32, 33, 34, 35, 36, 37, 39, 40, 41, 42, 44, 45, 46, 49, 50, 51, 52, 53, 54, 55, 56, 59, 60, 61, 63, 64, 65, 67, 68, 69, 70, 71, 72, 73, 74, 75, 76, 79, 82, 84, 85, 86, 87, 88, 89, 90, 93
heating, 10, 13, 26, 27, 30, 35, 45, 46, 50, 59, 60, 67, 69, 72, 73, 75, 78, 85, 86, 87, 88, 90
helium, 51, 53, 55
high pressure, 2, 19
high temperature, 19, 50
housing, 17
hydro, 11
hydrocarbons, 11
hydrogen, 51

I

inertia, 18
infinite, 63
initial state, 12, 86, 87, 88, 89, 90
interactions, 49, 67
internal combustion, vii, 1, 17, 37, 41, 43, 59, 75, 76, 77, 78, 79, 80, 90, 93
isothermal, 49, 51, 53, 63, 65, 67, 69, 70, 72, 73, 87, 88

L

law, 2, 20, 33, 38, 44, 50, 68
limitations, 75
linear, 41
lubrication, 5

M

methane, 43
missions, 59
motion, 17, 18

N

natural, 63
natural environment, 63
noise, 4, 18, 59

O

octane, 4

P

performance, vii, 4, 12, 59, 93
pollution, 5, 59
poor, 5
ports, 17
power, 5, 8, 11, 12, 13, 15, 17, 18, 24, 28, 33, 43, 50, 57, 75

Index

pressure, vii, 1, 4, 5, 6, 7, 8, 11, 12, 13, 14, 15, 16, 19, 21, 22, 23, 24, 25, 26, 27, 28, 29, 30, 31, 32, 34, 35, 36, 37, 39, 41, 42, 43, 45, 46, 51, 53, 55, 57, 60, 61, 62, 65, 67, 70, 72, 76, 79, 86, 87, 88, 89, 90, 93

R

range, 12, 63
reservoir, 50, 63
reservoirs, 52, 63
resistance, 4
returns, 86, 87

S

SAE, 59
sea-level, 57
sensitivity, 6, 7, 21, 22, 23, 39, 40, 77, 80
software, 75
spark-ignited, 17
specific heat, 3, 21, 34, 39, 44
speed, 53
stages, 17
standards, 18
stroke, 1, 2, 4, 5, 11, 13, 15, 17, 19, 28, 33, 43, 57, 58, 59
strokes, 1, 5, 11, 17, 59
substitutes, 17

T

temperature, 1, 4, 5, 6, 7, 8, 9, 10, 12, 13, 14, 15, 16, 19, 21, 22, 23, 24, 25, 26, 27, 28, 29, 30, 31, 32, 34, 35, 36, 39, 41, 42, 45, 46, 49, 50, 51, 53, 55, 59, 60, 61, 62, 63, 65, 67, 70, 72, 76, 79, 88, 89, 90
theory, 63
thermal efficiency, vii, 3, 7, 9, 10, 11, 13, 14, 15, 16, 20, 21, 29, 30, 31, 32, 34, 39, 44, 51, 59, 60, 61, 62, 69, 93
thermodynamic, 2, 63
thermodynamics, 2, 20, 33, 38, 44, 50, 68
time, 5, 17, 37, 57
timing, vii, 57, 59, 93
transfer, 37, 52
trucks, 28, 59

U

users, 75

V

values, 36
variable, vii, 59, 93
variation, 57
velocity, 17
vibration, 18

W

water, 63
weight ratio, 18